Maple® Laboratory Manual

for

CALCULUS: MODELING AND APPLICATION

Smith/Moore

Single Variable Labs

Richard M. Schori
Oregon State University

William H. Barker
Bowdoin College

David A. Smith
Lawrence C. Moore
Duke University

With the Assistance of Todd S. Coffey

D. C. Heath and Company
Lexington, Massachusetts Toronto

Address editorial correspondence to:
D. C. Heath and Company
125 Spring Street
Lexington, Massachusetts 02173

Trademark Acknowledgment: Maple is a registered trademark of Waterloo Maple Software.

Copyright © 1996 by D. C. Heath and Company.

All rights reserved. No part of this publication may be reproduced or transmitted in any form or by any means, electronic or mechanical, including photocopy, or any information storage or retrieval system, without permission in writing from the publisher.

Published simultaneously in Canada.

Printed in the United States of America.

International Standard Book Number: 0-669-32794-8

10 9 8 7 6 5 4 3 2 1

Preface

• Primary Use of the Manual.
This lab manual was designed as a supplement to the text *Calculus: Modeling and Application* by David A. Smith and Lawrence C. Moore of Duke University. It also can be used independently or with other texts with the understanding that certain labs will need some supplemental background.

• Contents of this Manual.
This manual contains printed copies of thirty-one *Maple* laboratory worksheets for first year calculus. These laboratory experiments are for the most part conversions of *Mathematica* notebooks by William H. Barker of Bowdoin College. Many of the original labs were written in *MathCAD* or *Derive* by the Project CALC team at Duke University.

• What is Project CALC?
Project CALC (Calculus As a Laboratory Course) is an award-winning curriculum development project funded by the National Science Foundation. It has produced materials for a three-semester calculus course that emphasizes real-world problems, hands-on activities, discovery learning, writing and revision of writing, teamwork, intelligent use of available tools, and high expectations of students.

• What is *Maple*?
Maple is a professional-level computer software package that combines text material with numerical and symbolic computations, equation solving, graphics, and a high level programming language. This combination of text, animated graphics, and "live" formulas gives a good example of what is now being called an *interactive text*.

• How to use the Worksheets?

We recommend you use these labs in a two-hour per week computer laboratory which accompanies a three lecture per week class schedule. These Laboratories use Maple V R3 in either the DOS/Window or the UNIX/X-Windows environment. You can use one of these labs per week or fewer, interspersed with other activities or labs designed at your home institution. Versions for Macintosh computers will be available soon after the release of Maple V R4.

We recommend that students work in groups of two or three per computer.

Students should read the printed copy of the laboratory prior to a lab session. In the actual lab students work through the electronic version answering questions, making conjectures, and analyzing results. Results are recorded directly into the computer file. Student groups make a second copy of the electronic worksheet and delete from it everything but the question/answer material and then submit the results to the instructor either electronically or as hard copy, or they may be asked to write a formal report.

• How Much *Maple* Background is Needed?
The Project CALC *Maple* worksheets are designed for students with some minimal computer background, but no *Maple* background. Lab 1 is designed to provide an introduction to *Maple* with additional help in Lab 2 and within other labs as needed. In general, these labs are designed to be "user friendly." Most of the labs have built-in computational and graphical commands to help students understand the basic ideas of calculus without being overwhelmed by the technicalities of *Maple*. There are many optional introductions to *Maple* if you want to skip the introductory Project CALC labs.

• How to Obtain Electronic Copies of the Worksheets?
Course instructors can obtain electronic copies of the *Maple* laboratory worksheets directly from the publisher, D.C. Heath and Company. Inquiries may be made about the possible existence of an ftp site for the UNIX versions.

• Hardware and Software Requirements.
The current *Maple* Project CALC laboratories use Maple V3 and were developed on a DOS/Windows platform. The labs should work well on any PC that is running Maple V3 for Windows with the usual suggestion of 8 MB or more of RAM and at least a VGA monitor. Versions for *Maple V4* will be available shortly after that version of *Maple* is released.

Slightly modified versions of the files are available for the version of *Maple* V R3 that runs under X-Window on a UNIX system. The author has used the UNIX version of the labs in teaching an Honors calculus class at Oregon State University.

• Acknowledgments.
The principal author of these *Maple* laboratories wishes to thank the Interactive Mathematics Text Project (IMTP) for introducing him to *Maple* as an authoring tool for interactive texts. The IMTP, under the leadership of Gerald J. Porter of the University of Pennsylvania, provided the numerous workshops and conferences, hardware, software, and inspiration that were a necessary part of this project.

We also thank Waterloo Maple Software for their support of the IMTP and this author by providing software and technical assistance, particularly from Stan Devitt and Benton Leong.

We thank Oregon State University for their indirect support of this project through an L.L. Stewart Faculty Development Award and the opportunity to use these *Maple* labs in teaching an Honors College calculus laboratory.

Finally, we thank Todd S. Coffey, Oregon State University undergraduate mathematics major, who has worked brilliantly and faithfully with the principal author for most of this project. Todd is not only an excellent computer person and programmer, but also knows mathematics, the pedagogy behind teaching mathematics, and has excellent judgment in dealing with complex projects. In many ways Todd could be viewed as a co-author, but because of the way the project developed, he is a person who very early in his career has the distinction of "Assisting with the authorship of these laboratories."

Contents

• **Notes to the Instructor.**
There is a great deal of flexibility in how you use Project CALC *Maple* labs. First, you probably don't want to use all of them. For example, Labs 12 A, B, and C, were designed to give you options and you should not be expect a student to do all of them. Also, if you are teaching a more traditional course, you may want to skip some of the labs that model physical or biological behavior. Examples are Lab 6 (Raindrops) and Lab 9 (The Spread of Epidemics). A good idea is to occasionally have your students do another lab project where the students start with a blank *Maple* screen.

As discussed in the Preface, we suggest that student groups of two or three work at a computer and also write reports as a group.

What do students turn in? There are basically two options: a short answer report, or a formal report, where the instructor exercises this option lab by lab.

To facilitate the short answer report, we often provide a Laboratory Results Summary Sheet at the end of the lab. At other times we ask a series of questions within the lab, for students to answer with short concise statements. In either case, student groups can be advised to make a second copy of the electronic version of the lab, and in it delete everything except the questions and answers. This short version is submitted electronically or as hard copy.

Writing a formal report on a lab is a wonderful learning experience and many universities are requiring a certain number of "Writing Intensive Courses." The full treatment is to have a *double submission* of the report. This means that the instructor corrects the first draft and the students rewrite it for a second submission.

• **Notes to the Student.**
Your main objective is to learn from these labs, and it helps to enjoy them and even become engrossed in them. Many students do. An interesting motto is "Talking isn't teaching, and listening isn't learning." A corresponding statement for the latter is that "Hitting the Enter key is not learning." We rarely let you proceed very far in a lab without needing to answer a question or complete a command. Sometimes we ask you to write answers on paper. The idea is that many of us learn best by actually writing things down in our own hand writing. As you work through these labs, keep thinking about how you can most efficiently enhance your understanding of the ideas.

• **Writing Formal Reports.**
We hope your instructor occasionally requires you to write a formal report on a lab. All formal laboratory reports for this course should include answers to the following general questions:

(1) What were the goals of the project?
(2) What actions were taken during the project?
(3) What were the observed results of these actions? and
(4) What conclusions did you draw from your observations and how do you justify them?

See Lab 2 for more suggestions and ideas on how to write a formal report. Most of the ideas would apply to writing a formal report on any of the labs.

LAB 1: Functions Defined by Data

□ NAMES:

□ PURPOSE.
The MAPLE Tutorial gives you enough information to begin working with MAPLE running under either DOS/Windows or X-Windows on a UNIX machine. The Project then uses MAPLE to explore three functions defined by data: the changing U.S. postage rates, the spread of AIDS, and the Federal budget deficit.

□ PRELIMINARIES.
Look at the shaded margin to the right (the scroll bar); at the top is an upward pointing arrow, while at the bottom is a downward pointing arrow. Using the mouse, move the cursor onto the down arrow and then "click" the mouse button. Now click on the up arrow. Now click on the down arrow and hold the button down for a few seconds.

You now are able to maneuver around this document. Faster movements are accomplished by clicking the mouse in the scroll bar just above or just below the scroll box (i.e., the small moving square inside the scroll bar). See what this does by trying it.

□ PREPARATION.
It is assumed that you have had a short introduction to the most basic aspects of your computer (i.e., What is the mouse? How does it move the cursor?). Most future labs will have a preliminary reading assignment from your text, Calculus: Modeling and Application.

□ LABORATORY REPORT INSTRUCTIONS.
The only "report" required for this lab is your electronically saved, completed notebook. Your instructor will tell you what should be done with your completed file.

□ MAPLE TUTORIAL.

¤ 1. THE WORKSHEET MODEL.

Maple V uses a "Worksheet" model which is arranged into regions. There are four distinct types of regions: text, input, output, and graphics. These regions may be separated by Separator Lines.

¤ Separator Lines.
The only significant place where the Windows and UNIX versions differ is in their handling of Separator lines. When there is a difference we will provide instructions for both.

[Windows]
If you don't see horizontal lines on your screen, under the
• View menu, click on Separator Lines, or hit
• F9 on your keyboard. The F9 key toggles back and forth between showing lines or not.

It is best to keep the Separator Lines visible. These lines separate regions or region groups in the worksheet.
• Create a new Separator Line above cursor --- F3 key.
• Remove Separator Line above cursor --- F4 key. Try these.

[UXIX]
If you don't see horizontal lines on your screen, under the
• View menu, click on Display Region Boundaries.
It is best to keep the Separator Lines visible. These lines separate regions or region groups in the worksheet.
• Create a new Separator Line below cursor --- Meta+S. (The Meta key is to the left of the space bar and has a triangle on it.)
• Remove Separator Line below cursor --- Meta+J. Try these.

¤ Input Lines.
Input lines are lines that begin with a ">" and their purpose is to contain Maple computer commands. You need to be able to change text lines into input lines and visa versa.

[Windows]
• F5 converts Maple text line to input line and converts an input line to a text line. Try it.

[UNIX]
• Meta+I creates an input region below the next Separator line. So Meta+S, up arrow, and Meta+I in that order creates an input line at the current position of the cursor. Or, if the cursor is in a text region, then the Text button in the Menu Bar can be used to convert the whole region to an input region.

• Maple Tips.
1. To "activate" or "Enter" a Maple command line means that you place the cursor anyplace on the line and press the Enter (or Return) key. Alternately you can double click the left button on the mouse.
2. The next line starting with a > is a Maple command. The command, 2 times 5, is followed by a semi-colon (;) which tells Maple that the command has ended. If you Enter a command line not ending with a semi-colon, the command is not executed. The best way to correct this is to put a semi-colon by itself on a command line immediately below and hit the Enter key. This will terminate the command and the output will be displayed. Then go back and fix up your original command. Enter the next command and experiment in the following line with no semi-colon.

```
> 2*5;
> 2*5
> ;
> # Scroll down with the down arrow.
```

• Maple Tips.
a) On an input line, Maple ignores everything after a #. This allows us to make comments on input or command lines.

Thus, in a Maple worksheet you can combine Maple Input, Maple Output, and Text comments in the same Worksheet. You can also paste graphics into the Worksheet.
b) When you Enter (or activate) an input line,
- output appear right before the next Separator Line, and the
- cursor jumps ahead to the next Input Line!!!

Yes, even if it is pages ahead. Go back and look at the last input line, the one containing the # and a helpful comment. Without it the cursor would have jumped ahead to the next input line which is a page or so below what you are reading now.

¤ 2. EDITING WORKSHEETS.

We will explain some basic editing procedures here and do more in Lab 2.

We move or delete items as follows.

- Select a portion of the text. Do this by placing the cursor at one end of the desired selection, click and hold the left mouse button down while dragging the cursor to the other end of the selection. Now release the mouse button and the desired selection should be highlighted.

From the Edit menu at the top of the worksheet you can Cut, Copy, Paste, and Delete any part of your Maple worksheet. Key board shortcuts are listed below in parentheses.
You start by highlighting the material in question.

- Cut (Ctrl+x) removes the highlighted material but keeps it (temporarily in memory) so that you can Paste it into another portion of the worksheet or into another Maple session on your same computer.
- Copy (Ctrl+c) puts the highlighted material into the computer memory for pasting elsewhere, but leave the existing material in place.
- Paste (Ctrl+v) puts the last Cut or Copied material into the worksheet at the current location of the cursor.
- Delete removes the highlighted material and it is gone, just as with the Delete key.

Or just start typing and the highlighted material is replaced by what you are typing.

Spend a little time practicing these editing procedures.

¤ 3. SAVING THE WORKSHEET.

Important! Have you Saved your work recently? If not, then open the File menu and choose Save. Keyboard shortcuts:

[Windows] --- Alt+f, s, or click left button on third (save) icon on Tool Bar.

[UNIX] --- Meta+f, s.

¤ 4. COMPUTATIONS WITH MAPLE.

Activate the following lines.

```
> a := 2^3*sin(Pi/4);
> b := evalf(a);
> # Scroll down with a down arrow.
```

The command

a := 2 means a has been assigned the value 2, whereas

a = 2 is a boolean expression used in programming situations like if a = 2, then ... .

The command

evalf(a) means the "numerical approximation" of a. For example

evalf(Pi) -> 3.14159265... .

Maple uses these standard symbols for operations.

+	addition
-	subtraction, negation
*	multiplication
/	division
^	exponentiation

Some commonly used functions are

sin(...), cos(...), tan(...), the trigonometric functions;

sqrt(...), the square root function;

Pi, the constant = circumference/diameter of a circle.

E, the constant e (base of the natural logarithm) where evalf(E)=2.71828182

Warning! You must use the exact notation given above, e.g., if you want e you must use E, not e, and for the sine of x, you must type sin(x), not Sin(x) nor Sin[x] nor sin[x] nor sin x, etc. Maple's demand for precise conformity to its notational rules is a source of headaches for the beginning user. Take care to develop correct notational habits now--it will save you hours of frustration later! Maple is case sensitive, so use Pi and not pi.

Exercise. Create and Enter input cells to numerically evaluate the following quantities:

sin(0.6), e^2, tan x when x=0.375, sqrt(2.3^2 + 1.1^3)

• Creating INPUT lines. We introduced this above but give more details here.
[Windows]
Start with a blank line.

1) Hit F5 to create an input >, but no Separator Line is created.
2) Move down one line and create a Separator Line with F3.
3) Move up to the input line and hit Ctrl+I, or Tab to create another input line.

[UNIX]
Start with a blank line.
1) Hit Meta+S to create a Separator Line below the cursor.
2) Move up one line and create and Input Line with Meta+I.
3) Use Meta+I again to create another Input Line right below the first one.

(As a check, your answers should be: 0.5646424, 7.389056, 0.3936265, 2.5731389)

If you get an error message when entering any of your expressions, then edit the expression to correct the problem and re-enter the line. Use evalf to get a numerical approximation for E^2.

¤ 5. THE ORDER OF EVALUATION.

A collection of input lines or regions has two different "orderings":

(1) the physical location of the lines, i.e., as you move from beginning to end in the worksheet; and

(2) the order of entry, i.e., the first line you entered, the second line you entered, etc.
Enter the following lines in the reverse order. (Put cursor on the line and hit the Enter key.)

```
> x := 1;    # Enter this line second.

> x := 2;    # Enter this line first.
```

Now find out what value Maple currently assigns to the symbol x. Enter the following line.

```
> x;
```

From this output, which ordering of the lines determined the value for x, the physical location of the lines or their order of entry?
Answer:

This section illustrates....

The Order of Entry Principle. The "physical location" of input lines does not determine the values assigned to variables. The order of entry of the lines determine the values.

Ignorance of this principle will lead to confusion and errors of magnificent proportions!

To restore x to a variable, activate the following line.

```
> x := 'x'; # The apostrophe is to the right of the semicolon key.
```

Now see what x is by entering the following line.

```
> x;
```

¤ 6. EXPRESSIONS WITH VARIABLES.

Suppose z depends on t. Here is an example. Enter the following definition for z in terms of t.

```
> z := 'z'; t := 't';
> z := t^2 + 1;
```

Suppose we wish to evaluate z when t=2. One method is to enter t :=2, followed by the statement z, as done below. Enter the following input line.

```
> t :=2; z;
```

We can recover the original (unevaluated) expression for z by removing the value given to t. Enter the following input line.

```
> t := 't'; z;
```

Suppose, however, that we wish to evaluate z when t =2 without actually changing the values of either t or z. This is done with the subs operator. Enter the following input line.

```
> subs(t=2,z);
```

The statement just entered means "substitute t equals 2 into z" The actual definitions of t and z remain unchanged, as will be seen by entering the next line.

```
> t; z;
```

Evaluating quantities by means of the substitution operator subs is a common Maple procedure. It seems like an odd construction when first encountered, but it proves to be very useful.

Practice. Use the substitution operator to determine the numerical value of w = 1/(x^2+1) when x equals the values 1, 2.5, Pi, and sin(0.6)+1. Create and evaluate your input lines below.

(As a check, your answers should be: 0.5, 0.137931, 0.0919997, 0.290014)

¤ 7. SAVING THE WORKSHEET AGAIN.

Have you Saved your work recently? If not, then open the File menu and choose Save.
• Save as... under the File menu: Select directory where you want your file stored, otherwise Save as... will save your work into an unexpected place.

Important: If you Quit Maple now, you will have all of your work intact, and reopening the worksheet will show all the changes and evaluations made up until this point. However, nothing will be in Maple's memory! For example, the quantity z from the previous section will have no values or expressions assigned to them. In order for z to have the assignment given previously, you must reenter the line defining z.

When reopening a worksheet, be sure to Enter any input regions that assign values to quantities needed for your subsequent work.

• Maple Tip. Before saving your file it is often convenient to remove all the output. To do that,
[Windows] under the Format menu, activate Remove All, and then Output.
[UNIX] under the Edit menu, activate Delete by Type, and then All Maple Output.

¤ 8. DEFINING FUNCTIONS.

You can define your own functions for use in computations. For example, the following defines the function f where f(x) = sin(2x). Enter the following region.

```
> x := 'x':              # Clears x of possible previous assignment.
> f :=  x -> sin(2*x);   # This defines the function f defined by f(x)=sin(2*x).
>
```

It is of critical importance that you know the rules for creating functions. You will need to create functions almost every time you use Maple. We now show you the "arrow" method of defining a function. There are others.

• Rules for defining functions with the ARROW method.
(1) It is best to clear the desired variable(s) first.
(2) Immediately after the function symbol, f here, use the := assignment symbol. In particular, use
f := ... and not f(x) := ... .
(3) Next use the arrow symbol; ->, a - (dash or minus sign) followed by > (greater than).
(4) Finally type the desired functional expression, such as sin(2*x), ending with a ; .

To evaluate f at a particular value of x, say x=2.5 , merely Enter the expression f(2.5). (This operation does not change the stored values for x nor f(x).) Evaluate f(x) at x=2.5 below.

```
> f(2.5);
```

¤ 9. GRAPHING.

• The plot Command

plot(f(x), x=a..b),

plots the function f over the interval [a,b].

• To get rid of the plot, under the File menu of the Plot, activate Exit.

[UNIX] Tip. After entering a plot command a "wire" framing of the picture appears. Drag it to a convenient location on the screen, say the lower left-hand corner, and press the left button of the mouse to make the plot visible.

Enter the next line.

```
> plot(f(x), x=0..Pi);
> # If you get "Warning in iris-plot: empty plot" , Enter f in last section.
```

• Maple Tip. In UNIX,

• Coordinates of a point on a Maple plot can be displayed by clicking the mouse at the point. The coordinates will appear in the lower left part of the plot.

Practice. Define g(x) to be the function g := x -> 2*x^2 - 2*sin(x) - 1. Approximate the roots of this function by plotting the function over a variety of intervals of x values. Create and evaluate your function definition and plot statements below.

```
> g := x -> ??? ;
> plot(g(x), x= ???);
>
```

(As a check, one of your roots will be x=1.19608. Find the other.)

¤ 10. HOW MUCH MORE CAN MAPLE DO?

Much more--we have barely scraped the surface of this large program. However, you now have enough knowledge to perform some basic computations. Your knowledge will be used--and further extended--in the Project section which follows.
More about Maple will be given in subsequent labs. However, our philosophy is to discuss only as much about the program as we need for the tasks at hand.
Further introductory information about Maple can be found in books such as:

- Wade Ellis, Eugene Johnson, Ed Lodi, Daniel Schwalbe, Maple V Flight Manual, Brooks/Cole, 1992
- Bruce Char, et al, Maple V, First Leaves: A Tutorial Introduction to Maple V. Springer-Verlag, 1992
- Andre Heck, Introduction to Maple, Springer-Verlag, 1993
- John S. Devitt, Calculus with Maple V, Brooks/Cole, 1993

• Save the Worksheet again....
Have you Saved your work recently? If not, then open the File menu and choose Save or use a keyboard shortcut.

□ PROJECT.

¤ 1. Postal Rates.

The following is a list of first class postage rates (in cents) for the past century. Each rate is paired with the year in which it took effect. Enter this data set.

```
> postage := [[1885, 2], [1917, 3], [1919, 2], [1932, 3], [1958, 4], [1963, 5], [1968, 6],
> [1971, 8], [1974, 10], [1976, 13], [1978, 15], [1981, 18], [1982, 20], [1985, 22],
> [1988, 25], [1991, 29]];
>
```

• Maple Tip. If you don't want the data printed when entering a data set, add a colon to the end of the input cell. This suppresses the display of the output.

Each data point is a pair of numbers (year and postage rate) enclosed within square brackets [ ], and the full list of data points is itself enclosed in an outer set of square brackets [ ]. We call this a "list of lists" and it is in a good format for plotting data. We create a scatter plot of the collection of data by using the Maple plot command with the style=POINT option. Enter the following plot command.

```
> plot(postage, style=POINT);
```

See what happens without the style=POINT option by Entering the next line.

```
> plot(postage);
```

With the style=POINT option, if you want larger points, add the option symbol=CIRCLE.

```
> plot(postage, style=POINT, symbol=CIRCLE);
>
```

Have you been exiting the plots properly? Use the File menu of the Plot and activate Exit.

Write a brief description of what you learn from the graph about postage costs in the United States. Type your comments below.
Answer:

• ON MATHEMATICAL WRITING. Your response to this question--and to any similar questions in future labs--should be given in a coherent paragraph comprised of clear and complete sentences. Often one sentence will do--long responses are not automatically better than short ones. Your primary goal for a written response should be clarity. Ask yourself "Would somebody with knowledge and background similar to my own be able to understand what I have written?"

¤ 2. AIDS Cases.

The list AIDS, given in a region below, contains the total numbers of AIDS cases reported to the Centers for Disease Control as of the indicated dates.

Enter the AIDS data.

```
> AIDS := [[1981.67, 110], [1981.75, 129], [1982.00, 220], [1982.17, 257],
> [1982.42, 439], [1982.58, 514], [1982.92, 878], [1983.08, 1029], [1983.50, 1756],
> [1983.67, 2057], [1984.08, 3512], [1984.33, 4115], [1984.92, 7025], [1985.17,  8229],
> [1985.75,  14049], [1986.08,16458], [1986.92, 28098], [1988.17, 56575], [1989.92, 113891],
>  [1992.58, 226252]];
```

Plot the AIDS data by completing and entering the following command.

```
> plot(???);
>
```

(This data was derived from the "Morbidity and Mortality Weekly Report" of the Centers for Disease Control, supplemented with additional data obtained from Duke University.)

Create a plot of this data with the points joined together. (Reminder: The region containing the AIDS data must be Entered before it can be used in a plotting command.)

```
> plot( ??? );
>
```

Write a brief paragraph describing the AIDS epidemic graph. What sort of formula do you think might have a graph that fits this data? What can you learn from the table or the graph that you didn't know before? (Later in the course we will develop tools to perform a more rigorous analysis of a data set such as this.) Place your answers below.

Answer:

The fractional dates indicate that each report is associated with a particular month. Assuming that reports are filed at the beginning of a month, fill in the months for the last four reports: Place your answers below.

1986, _______
1988, _______
1989, _______
1992, _______

□ MORE ON MAPLE.

¤ Creating lists of data.

• sequence command:

seq(n^2, n=1..5) -> 1, 4, 9, 16, 25

Suppose we want a list of the first twenty squares of integers. We could type them in, but Maple can make the list for us much faster. Enter the following command.

```
> squares := seq( n^2, n=1..20);
```

This says "make a sequence (or list) of values of n^2, for values of n going from 1 to 20." Suppose we want to start with 0, and go only to the square of 5. We then simply change the limits for n. Enter the following command.

```
> n;  # The seq command from above leaves n assigned as follows.
> n := 'n':  # It is important to clear n here.
> someSquares := seq( n^2, n=0..5);
```

Suppose you want to enclose the list in [ ] brackets.

```
> n := 'n':
> someSquares :=[ seq( n^2, n=0..5) ];  # Enclose seq command in [ ].
```

To plot someSquares in a meaningful way, we should create a list of lists where, for each n, n is paired with n^2, i.e. [n, n^2]. Enter the following line.

```
> n := 'n':  someSquaresData := [ seq( [n,n^2], n=0..5)];
```

Plot someSquaresData with points replaced with circles by completing the following plot command. Complete and Enter the following command.

```
> plot(???);
>
```

Practice. Define someSines to be a list of lists consisting of ordered pairs of the numerical values [k/10, sin(k/10)] at all the points k/10 as k ranges over the integer values between 0 and 64. Plot the resulting list. Enter your commands below. Hint: If you use k/10. (10 with a decimal point) instead of k/10 , then Maple will do all the computations with decimal values.

```
> someSines :=[ seq([???  )];
> plot(someSines, ???);
>
```

(The plot should be quite simple and recognizable. If not, you have made a mistake....)

□ PROJECT continued.

¤ 3. Federal Budget Deficits.

A quick economics lesson. Since the early seventies, each year the Federal government has spent more money than it has received--the yearly shortfall in money spent compared to money received is called the (annual) Federal budget deficit. The following is information about Federal budget deficits over a recent 15-year period. The first number is the year, the second number is the budget deficit (in billions of dollars) that was amassed during that year. (For example, during the period January 1, 1977, through December 31, 1977, the Federal government made 53.6 billion dollars more in expenditures than it received in revenues.)

1977	53.6
1978	59.2
1979	40.2
1980	73.8
1981	78.9
1982	127.9
1983	207.8
1984	185.3
1985	212.3
1986	221.2
1987	149.7
1988	155.1
1989	153.4
1990	220.4
1991	318.1 (estimate)

(This data is from "Historical Tables, Budget of the United States Government, Fiscal Year 1992," p. 17.)

We now construct a data set deficit (much like postage and AIDS) which contains the information from this table. Enter the following data set command.

```
> deficit := [[1977, 53.6], [1978, 59.2], [1979, 40.2], [1980, 73.8], [1981, 78.9],
> [1982, 127.9], [1983, 207.8], [1984, 185.3], [1985, 212.3], [1986, 221.2],
> [1987, 149.7], [1988, 155.1], [1989, 153.4], [1990, 220.4], [1991, 318.1]];
```

Plot the data below.

```
> plot( ???);
>
```

Write a brief description in words of what you see in the deficit graph. What do you learn from the table or the graph that you did not know before? Type your response below.

More economics. If you spend more money than you make, you eventually go into debt, i.e., you will have to borrow money from some source in order to pay your bills. The Federal government is no different. Because of the number and size of the yearly deficits during this past century, the Federal government is up to its eyeballs in debt. The total Federal debt as of January 1, 1980 was about 900 billion dollars. Use this, and the data on the budget deficits during the 1980's and early 90's, to estimate the total Federal debt at the start of 1992. Type and Enter your calculations below.

What was the percentage increase in the debt over this 12-year period (i.e., the amount of debt added during this nine year period divided by the total debt at the start of the period)? Type and Enter your calculations below.

Even more economics. If you go into debt and borrow money, the institution which lends you the money will charge interest. Interest, computed as a percentage of the amount of money you currently owe on your loan, is a fee paid to the lender on a regular schedule such as monthly or yearly. The Federal government is no different: it must pay interest on its huge Federal debt. If interest paid on the debt in 1992 averages 8%, about how much money had to be allocated in the 1992 Federal budget just for interest payments on the debt which existed as of January 1,1992? Type and Enter your calculations below.

Given a United States population of approximately 250 million people, how much interest was paid per capita in 1992? Said more descriptively, if the Federal debt interest payment in 1992 was distributed evenly over every man, woman, and child in the United States, how much money would each person owe? Type and Enter your calculations below.

Some critical thinking. An interviewee on National Public Radio (Sept. 5, 1992) claimed that the Federal budget deficit was currently 1 million dollars per day. Is that statement consistent with the data in this lab? If so, how do you figure it out? If not, what might explain the discrepancy? Type and Enter your calculations and conclusions below.

- Save the Worksheet again....

- Printing the Report.

If you wish, you can also make a printed paper copy of the Worksheet for yourself on the laboratory printer. Instructions on how to use the printer for this purpose will be supplied by your instructor.

- Quitting.

First open the File menu and choose Save, then reopen the File menu and choose Exit.

LAB 2: An Example of Marine Pollution

□ NAMES:

□ PURPOSE.
The major aims of this laboratory are to gain familiarity with further plotting and text-processing aspects of Maple and to write a short paper focused on data concerned with marine pollution.

□ PREPARATION.
You should read the first two sections of Chapter 1 in Calculus: Modeling and Application.

□ LABORATORY REPORT:
Disregard this if instructions from your lab instructor are different.

You are to write a self-contained exposition and analysis of the marine pollution study discussed in this laboratory. In particular, it should incorporate the answers to all the important questions raised during the lab session.

Two submissions will be made for this report: an initial draft, and a final revision. Comments made on the draft will help you prepare the revision, and only the revision will receive a grade. More specific instructions on the content of the report will be given at the end of the notebook. As you will see, the point-of-view we wish to take is that of a science reporter for the statewide newspaper The Maine Times.

Maple Procedures. This lab, like most of these Project CALC labs, has some "behind the scenes" programming providing additional Maple commands to help the student. Please Enter the following line to "initialize" this lab.

```
> with(lab2);
> # If an Error is an output of the above line, see your lab instructor.
```

□ PROJECT.

¤ 1. The Effects of TBT Pollution on Mussels.

Tributyltin (TBT) is a chemical which is acutely toxic to many species of marine animals. For this reason it effectively inhibits the growth of marine organisms when used in marine paints (e.g., on boat hulls) and in other "antifoulants." However, concern about the adverse effects on marine life of TBT leaching from treated surfaces has resulted in strict controls of its use in France, the United Kingdom, the US, and other jurisdictions.

While Maine, a state with a low population density, is perceived as having a relatively unpolluted coastline, there are activities that do pose pollution threats. In the case of butyltin species, activities such as shipyards, pleasure craft, commercial ship traffic, and wood and fishing net preserving are all potential sources of TBT. Because of this, members of the Bowdoin College

Hydrocarbon Research Center and the Marine Research Laboratory conducted a study (1987-1989) to (1) learn about the extent of TBT inputs in Maine locations; (2) learn about TBT in mussels from various field locations; and (3) relate TBT in mussel tissues to an overall indicator of stress. (Mussels were used because they are an effective organism for detection of marine pollutants.)

Collections of blue mussels, Mytilus edulis, were sampled from 15 coastal Maine intertidal locations (plus one location in England) at various times between October 1987 and November, 1989. Six of the samples in 1987 formed three pairs, with one member of each pair in an area of high marine activity and the other in the same general vicinity but more remote from such activity.

As part of this study, David S. Page, Professor of Chemistry at Bowdoin, and Tamara Dassanayake, Bowdoin Class of 1990, analyzed the relationship between mussel shell thickness (as an indicator of stress) and the concentration of TBT found in the cell tissues of the organisms.

Shell thickness index (TI) was determined on the shells of 12-25 animals collected from each site (TI is defined to be 100 times shell depth, i.e., "thickness", divided by shell length. The concentration of TBT in the cell tissues was measured in micrograms (ug) per gram of tissue (the measurement of this quantity required a complicated analytical procedure whose description we omit). The results are given below.

Results of TBT and Thickness Index analyses in the tissue samples of mussels

ug/g TBT	Thickness Index
.04	.40
.11	.42
.50	.43
.60	.43
.73	.46
.77	.46
1.75	.48
4.31	.53
1.66	.50
3.56	.56
.27	.48
.45	.45
.74	.44
1.75	.48
.04	.42
.04	.39

In order to examine this data, we will place it all into a Maple list, simply a collection of data pairs using square brackets [ ] to demark the pairs and the full collection itself. Complete and Enter the following line.

```
> musselData := [ [.04, .40], [??? ... ]];
>
```

• Maple Tips.

a) Note that the := is used and not just the = sign. Also, be sure you end the above Maple command with a semicolon ; (or a colon : if you do not want to see the output) and Enter this input statement!

b) A scatter plot example. Now use the plot command to generate a scatter plot of the data. (If you don't remember how, check with Lab 1 or at the command cursor > type ?plot for a Help File. Or, this may help: if data := [[1,2], [3,4]], then plot(data, style=POINT) plots the two points (1,2) and (3,4) in the plane. Type and Enter your plot command below:

```
> plot( ??? );
> # Scroll down with the down arrow.
```

• Maple Tips.

[Windows]

c) To control the size of plots, place the cursor anywhere on the side of the graph until a double ended arrow appears, click down and hold the left button on the mouse, and drag it in some direction, and release the mouse. This adjusts the size of the graph.

d) To move the plot around on the screen, place cursor in bar at the top of the graph, click and hold the left button and drag the whole plot around!

e) To maximize the size of the plot, click on the up arrow in the upper right-hand corner of the plot with the left button of your mouse or hit Alt+Spacebar, x.

f) To exit the plot with the mouse, on the File menu click down the left mouse button, drag it down to Exit, and release it. Or, from the keyboard, hit Alt-F4.

[UNIX] After Entering the plot command a "wire" framing of the plot appears. Drag it to a convenient location on the screen, say the lower left-hand corner, and press the left button of the mouse to make the plot visible.

c') To control the size of plots, place the cursor on the small shaded square in the upper right corner of the plot. With the left button on the mouse drag the cursor around changing the size of the plot.

d') To move the plot around on the screen, place cursor in bar at the top of the graph, click and hold the middle button, and drag the whole plot around.

e') To maximize the size of the plot, move it to the lower left corner of the screen with d', then enlarge it to the whole screen size with c'.

f ') To exit the plot with the mouse, on the File menu click down the left mouse button, drag it to Exit, and release it. Or, from the keyboard, hit q or Meta+x.

g) Plot "options" enhance your graphs. Some general examples and specific recommendations for the above musselData plot.

You can specify the x and y ranges.

x range: x = xmin..xmax, x = -1.2..4.5 (x varies from -1.2 to 4.5)

y range: y = ymin..ymax, y = .33.. .585 (y varies from .33 to .585)

plotting points: style = POINT (The default is LINE for functions.)
symbol = CIRCLE (Or BOX, CROSS, or DIAMOND.)
color: color = red (Or green, blue, yellow, violet, etc.) or for more control,
color = COLOR(RGB, a, b, c) where each of a, b, c is a decimal between 0 and 1, specifying how much Red, Green, and Blue you want. For example COLOR(RGB, 0.1960, 0.6000, 0.8000) is sky-blue. See ?plot,color for more details.
axes labels: labels = [`ug/g TBT`, `Thickness Index`].
(Labels for x and y axes. Note backward quote, often found above the Tab key.)
The general format for plot options is
plot(data, option1, option2, option3, ...);
See ?plot, ?plot,options for more details.

Complete and Enter your enhanced plot command below.

```
> plot(musselData, ???);
```

• Maple Tips.
h) If you assign a name to a plot, like musselPlot, and end the command with a ; then on Entering the command only the technical graphing data is shown as output and not the plot. Try it. It is preferable to end the command with either a : or :"; . Ending with a : puts the plot in memory under the designated name but with no immediate output, while ending with :"; stores and displays the graph immediately.

```
> musselPlot := plot(musselData, x=-1.2..4.5, y=.33.. .585, style=POINT,
>    symbol = CIRCLE, labels = [`ug/g TBT`, `Thickness Index`], color = red):";
```

i) A good reason for giving the plot a name and putting it in memory with either the : or :"; is that later (in the same session) you can recall the plot (and possibly others at the same time) with the plots[display] command.

```
> plots[display](musselPlot);
> # Scroll down with the down arrow.
```

What sort of relationship seems to exist between the concentration of TBT in the cells of the mussels and the thickness of their shells?
Type response here:

Suppose we hypothesize that a linear relationship exists between the concentration of TBT in the mussels and the thickness of their shells. As a rough test of this hypothesis we can use Maple to help determine the "best linear approximation" to the data, i.e., the line whose graph seems to most coincide with the data points. A general line has the equation:
y = a*x + b,
where a is the slope of the line and b is the y-intercept. We assign temporary values to a and b and plot the line.

```
> a := 1: b:= 0:
> plot(a*x+b, x=-1.2..4.5);
>
```

The next Maple commands allows us to plot musselPlot and linePlot together with any desired values of a and b. You should read it as "Plot, in one graph, the scatter plot of our data and the graph of the function a*x + b (where you can assign different values to a and b) as x ranges from -1.2 to 4.5." Activate the following commands. See ?plots,display .

```
> a := 1: b := 0:
> linePlot := plot(a*x+b, x=-1.2..4.5, color=blue):
> plots[display]({musselPlot, linePlot}); # Maximize the plot.
>
```

• Maple Tip.

j) The plots[display] command used with several plots, say plot1, plot2, and plot3, will plot them all on the same graph. Warning! You need to enclose the set of plots in either { } brackets or [ ] brackets. When using { } you don't know in which order the graphs will be plotted. If you use [ ], the graphs are plotted in the order presented.

Does the line give a good approximation to the data? If not, then change the values for a and b and activate the statement again. Keep trying values of a and b until you feel you have obtained a good linear approximation to the data. Now, copy and paste this graph into this worksheet.

• Maple Tip.

k) Maple has built in routines for fitting specified types of curves (like a line) to a data set. In the "stats" package is a "fit" package that has a command called "leastsquare". This finds a line that best fits a data set and is called the "least square" linear approximation. To simplify the use of this "leastsquare" command, a command called "fitLine" has been written for this lab.

```
> fitLine(musselData);
```

Now, let's include the graph of this with the other two graphs. Type in the right hand side of the above equation into the fitPlot command below and activate it.

```
> fitPlot := plot(???):";
```

Now activate the following command to see all three graphs together, and see how close you came. Note that the plots are enclosed in {} brackets.

```
> plots[display]({musselPlot, linePlot, fitPlot});
>
```

You will write a report at the end of this Worksheet discussing the implications of the results obtained in this section. First, however, we explore some aspects of word-processing in Maple that will be helpful in producing written reports.

• Maple Tips.

[Windows]
l) On each Maple Plot, the top bar will say "Plot for Maple V - (some integer)". If the integer is not a 1 , then you are holding some Plots in computer memory. This happens when you get rid of a Plot by clicking on the worksheet, which puts the worksheet on top. Too many hidden plots will eat up your computer's memory. Find out-of-sight Plots in memory with the Alt+Tab combination. Now from the Plot's File menu click on Exit or hit the Alt-F4 keyboard shortcut.

[UNIX]
l') Maple plots under UNIX are not numbered. With a Maple plot on top of a Maple session, clicking the left button of the mouse on the top bar of the Maple session will cause the plot to disappear from sight. However it is using up memory and you should not keep many plots hidden in this way. Shift + right button will bring hidden windows and plots to the surface. The easiest way to kill a Maple plot is to hit q . In general, placing the little arrow anywhere on a partially covered window and using Shift + left button will bring that whole window to the top, and Shift + right button will bring hidden windows to the top.

¤ 2. Saving the Worksheet.

IMPORTANT: You should periodically save your work by opening the File menu and choosing Save, or

[Windows] click on the third icon in the "Tool Bar", (if it is not showing, under the View menu, check the Tool Bar), or hit Alt+f, s.

[UNIX] Use the Meta+f, s combination.

¤ 3. Text Editing Procedures.

In this section we consider some additional text editing features of Maple. This continues the discussion begun in the previous lab. Here is a sample text paragraph with which you will work:

"It is simplistic to suppose that people remember what they are told, and understand the things that are explained to them clearly. More commonly, people remember what interests them, and understand the things that they enjoy understanding." Edwin Moise.

a. Highlighting Text.

• Click, Drag, and Release • the left button of the mouse to highlight areas of the worksheet. Hit the Delete key, and the highlighted text will disappear, or start typing and the highlighted word disappears.

Exercise.

[Windows]
Click on the beginning of the first occurrence of "people" in the above quote and drag the cursor down a line or two before releasing the button. What happened? Does the whole line that contains the word people become highlighted? It does in Maple for Windows, Release 3. How could you highlight the material starting with people for the rest of the quote?

[UNIX]
Click on the beginning of the first occurrence of "people" in the above quote and drag the cursor down several lines before releasing it. What happened? Now insert a separator line below the quote (Meta+S) and repeat the highlighting process bringing the cursor down below the separator line. What happens?

• Shift key.
Hold down the Shift key to keep highlighted text highlighted while you move the little arrow further down the page prior to clicking the left button of the mouse. Also, with the shift key held down you can hit the down arrow on the keyboard to extend the highlighted area.

"Your learning is useless to you till you have lost your text-books, burnt your lecture notes, and forgotten the minutiae which you learnt by heart for the examination. What, in the way of detail you continually require will stick in your memory as obvious facts like the sun and the moon; and what you casually require can be looked up in any work of reference. [The function of education] is to enable you to shed details in favor of principles." Alfred North Whitehead.

b. Edit: Cut, Copy, Paste, and Delete.

From the Edit menu at the top of this worksheet you can Cut, Copy, Paste, and Delete any part of your Maple worksheet. Keyboard shortcuts are listed below. Start by highlighting the material in question.

• Cut (Ctrl+x) removes the highlighted material but keeps it (temporarily in memory) so that you can Paste it into another portion of the worksheet or into another Window document. (In MapleV Release 3 you can only move one line at a time from the worksheet into Notepad.)

• Copy (Ctrl+c) puts the highlighted material into the computer memory for pasting elsewhere, but leave the existing material in place.

• Paste (Ctrl+v) puts the last Cut or Copied material into the worksheet at the current location of the cursor.

• Delete removes the highlighted material and it is gone, just as with the Delete key.

• [UNIX] Delete Cursor Region (Ctrl+Del) removes the entire region the cursor is in.

c. Changing Fonts for text, input, and output regions.

[Windows]
Under the Format menu go to Fonts and then in turn to Input, Output, and Text. Experiment with changing the Font, Font Style, Size, and Color. A larger size may make it much easier to see, but less will fit on the screen.

[UNIX]
Under the Options menu go to Change Fonts and then in turn to Input, Text, and Output Fonts. Experiment with changing the Typeface, Style, and Size. A larger size may make it much easier to see, but less will fit on the screen.

□ LABORATORY REPORT--Specific Instructions.

As the science reporter for the statewide newspaper The Maine Times, you have been assigned to write a short article concerning the Bowdoin College study of the effects of TBT pollution on mussel shell thickness along the Maine coast. This news report will be your laboratory report.

You may assume your readers are mathematically literate (yes, we know that is a questionable assumption!)-- hence you can include formulas, such as the equation for a straight line, or data plots. However, you cannot assume that your readers have been through any part of this laboratory exercise--your report must therefore be self-contained, i.e., it must be understandable as a stand-alone essay, independent of the rest of the Worksheet. You should also decide on a point-of-view for the article, e.g., I must warn my readers of the danger of TBT pollution as revealed by the study, or I believe the study shows that the dangers of TBT pollution are not severe, or The study does not present enough data to form a definitive conclusion concerning TBT pollution. (Other points-of-view are also possible.)

All laboratory reports for this course should include answers to the following general questions:
(1) What were the goals of the project?,
(2) What actions were taken during the project?,
(3) What were the observed results of these actions?, and
(4) What conclusions did you draw from your observations and how do you justify them?

Specific questions that you could address in your article include:
(1) Does the data presented here convince you that a relationship exists between TBT concentrations in mussel tissue and the thickness index of their shells?,
(2) Does the data presented here seem to support or refute the hypothesis that a linear relationship exists between the two variables?,
(3) Could the variation of shell thickness be a result of factors other than TBT pollution?
(4) Should concern about TBT pollution be heighten or lessened by the results reported here? In particular, why should we be concerned about a few thick-skinned mussels?

This list of questions is designed to be suggestive, not definitive you might not choose to address every question we have listed, and you might choose to address questions that are not included on our list. In fact, we very much hope this happens! "Real world" uses of mathematics involve more than "answering questions" they involve determining the questions to be asked! Determining the questions, and interpreting the answers, is often of far more importance than the computational procedures themselves.

Important Qualification: The purpose of this written exercise is to have you think about the issues involved in examining data--you do not yet have the tools in hand to give precise answers to the posed questions. Rigorous analysis of data requires statistical techniques which are not covered in this course.

Your report should be entered at the location indicated below.

"Production Tip." One good way to proceed is to print a copy of your completed project for reference as you write the report (otherwise you'll be scrolling back and forth endlessly in this Worksheet).

Place your report below:

LAB 3A: Compound Interest

□ NAMES:

□ PURPOSE.
In this lab we explore questions related to the calculation of interest. We have three goals for this lab:
(1) To examine a simple case of exponential growth;
(2) to explore a particular type of limiting behavior; and
(3) to learn additional features of Maple.

□ PREPARATION.
Read Chapter 1 of Calculus: Modeling and Application. This provides general information concerning functions and relationships between two variables.

□ LABORATORY REPORT INSTRUCTIONS.
Follow you instructor's instructions if they differ from ours.

For this project you are not required to submit a formal report. Rather you are asked to submit an electronic copy of this Worksheet with the questions answered in the indicated locations of those portions of the lab that have been assigned. You will be graded on this single submission. Please write in complete sentences.

□ PROJECT.

¤ 1. A Patriotic Story of Interest.

One of your ancestors, a lawyer, had been in Philadelphia on July 4, 1776. He became so enthusiastic about the prospects for the new nation and his hopes for its long existence that he invested $1 with a local banker to gain interest at the rate of 3% per year, compounded annually, until July 4, 2000. At that time the entire sum was to go to one of his descendants. The particular formula for who is to receive this money is intricate, as befits a lawyer, but the upshot is that you are the lucky one.

What is your initial guess as to how much money will be in the account in the year 2000?

Place your guess here: ____________________

¤ 2. How much money will you receive in 2000?

Let's assign symbols to the important quantities in the problem:
k = the number of years the account has been in existence (on any chosen July 4);
A(k) = the amount of money in the account on July 4, k years after 1776;

We first need to record the value of A(0), i.e., the amount of money on July 4, 1776.
Place your answer below.

```
> A(0) := ??? ;
> # Scroll down with the down arrow.
```

As a preliminary task, we compute the values A(0), A(1), A(2), A(3), ... These will be the dollar values of the account during the first few years of its existence. Place these values in the following list (remember that the increase in each successive A value will be three percent of the previous value; in other words, 103% of the previous value!). Complete the following list.

In the year 1776.........A(0) = ---
In the year 1777.........A(1) = ---
In the year 1778.........A(2) = ---
In the year 1779.........A(3) = ---

Maple Tip. Be sure to write the interest rate as a decimal value, i.e., .03, not 3/100, the decimal point tells Maple to use numerical approximations.

From the computations, you should see a pattern that gives a formula for A(k), the account's dollar value in the year 1776 plus k years. Place your formula for A(k) below.

In the year 1776+k ...... A(k) = ---

Use your formula to express A as a Maple function of k. Complete and Enter the following function definition statement.

```
> A := k -> ??? ;
> # Scroll down with the down arrow.
```

To determine the k value in the year 2000, we must solve the equation:

$1776 + k = 2000$

We see that k will be 224 . You can now compute the account's dollar value in the year 2000.
Fill in your answer below.

In the year 2000.........A(---) = ---

Compute a numerical value for this quantity with Maple. Place your computations below.

If the output was A(224) instead of a number, then you didn't properly Enter the function A above. Does this value surprise you, i.e., was your conjecture from part 1 far off? If so, why?
Place your answer below.

¤ 3. What happens if the initial investment is doubled?

Suppose your ancestor had been twice as enthusiastic and had invested $2. What do you guess your account would be worth in 2000?

Place your conjecture here: ________________

In order to compute the precise final balance if the initial investment is doubled, let's follow the same procedures as in Section 2. We must record the value for A(0), then compute the values of A(1), A(2), A(3), ... These will be the dollar values of the account during the first few years of its existence. Complete the following list.

In the year 1776.........A(0) = ---
In the year 1777.........A(1) = ---
In the year 1778.........A(2) = ---
In the year 1779.........A(3) = ---

Do you see a general pattern emerging? Give a general formula to compute A as a function of k. Complete the following function definition statement.

```
> A := 'A':  k := 'k':            #Clear  A  and k.
> A := k -> ???
>
```

Use your formula to compute the value of the account in the year 2000. Enter your computations below.

Does this value surprise you, i.e., was your conjecture from part 1 far off? If so, why? Place your answer below.

¤ 4. What happens if the interest rate is doubled?

Suppose your ancestor had only invested $1, but had obtained an interest rate of 6%. What do you guess your account would be worth in 2000?

Place your conjecture here: ________________

In order to compute the precise final balance if the initial investment is doubled, let's again follow the same path as in Section 2. We must record the value for A(0), then compute the values of A(1), A(2), A(3), ... These will be the dollar values of the account during the first few years of its existence. Fill in the appropriate values below.

In the year 1776.........A(0) = ---
In the year 1777.........A(1) = ---
In the year 1778.........A(2) = ---
In the year 1779.........A(3) = ---

Do you see a general pattern emerging? Give a general formula to compute A as a function of k. Complete the following function definition statement.

```
> A := 'A':  k := 'k':
> A := k -> ??? ;
>
```

Use your formula to compute the value of the account in the year 2000. Enter your computations below.

Does this value surprise you, i.e., was your conjecture from part 1 far off? If so, why? Place your answer below.
Answer:

¤ 5. Why the big difference between Sections 3 and 4?

Write a paragraph to explain how and why the answers to Parts 3 and 4 are so different. Place your answer below.
Answer:

¤ 6. What happens if the interest is compounded twice a year?

Suppose your ancestor had only invested $1 at an interest rate of 3%, but that this interest was to be compounded twice a year, i.e., 1.5% interest on the balance of the account would be awarded each July 4 and January 4. What do you guess your account would be worth in 2000?

Place your conjecture here: ________________

In order to determine this number precisely, it is convenient to slightly alter the meanings of k and A(k). The new definitions will be as follows:

k = the number of interest periods which have gone by (on any given date);
A(k) = the amount of money in the account at the end of the k-th interest period.

In what ways do these definitions differ from the previous versions? Place your answer below.
Answer:

In order to determine the value if the initial investment is compounded twice a year, let's follow a similar procedure as in Section 2. We must record the value for A(0), then compute the values for A(1), A(2), A(3), ... These will be the dollar values of the account during the first few interest periods of its existence. Fill in the appropriate values below.

On July 4, in the year 1776.........A(0) = ---
On Jan 4, in the year 1777.........A(1) = ---
On July 4, in the year 1777.........A(2) = ---
On Jan 4, in the year 1778.........A(3) = ---

Do you see the general pattern emerging? Complete the following statement.

In the kth investment period... A(k) = ---

Use your formula to define A as a Maple function of k. Complete the following function definition statement.

```
> A := 'A': k := 'k':
> A := k -> ??? ;
>
```

Now determine the k for the year 2000, i.e., the number of investment periods which will have occurred by July 4, 2000. Place your answer below.

Use your formula to compute the value of the account in the year 2000. Enter your computations below.

Does this value surprise you, i.e., was your conjecture from part 1 far off? If so, why? Place your answer below.
Answer:

¤ 7. What happens if the interest is compounded three times a year?

Suppose your ancestor had only invested $1 at an interest rate of 3%, but that this interest was to be compounded three times a year, i.e., 1% interest on the balance of the account would be awarded each July 4, November 4, and March 4. What do you guess your account would be worth in 2000?

Place your conjecture here: ________________

In order to determine the value if the initial investment is compounded three times a year, let's follow a similar procedure as in Section 2. We must record the value for A(0), then compute the values for A(1), A(2), A(3), ... These will be the dollar values of the account during the first few interest periods of its existence. Fill in the appropriate values below.

On July 4, in the year 1776.........A(0) = ---
On Nov 4, in the year 1776.........A(1) = ---
On Mar 4, in the year 1777.........A(2) = ---
On July 4, in the year 1777.........A(3) = ---

Do you see the general pattern emerging? Complete the following statement.

In the kth investment period...A(k) = ---

Use your formula to define A as a Maple function of k. Complete the following function definition statement.

```
> A := 'A':  k := 'k':
> A := k -> ??? ;
>
```

Now determine k for the year 2000, i.e., the number of investment periods which will have occurred by July 4, 2000. Place your answer below.

Use your formula to compute the value of the account in the year 2000. Enter your computations below.

Does this value surprise you, i.e., was your conjecture from part 1 far off? If so, why? Place your answer below.
Answer:

¤ 8. What happens if the interest is compounded m times a year?

Suppose your ancestor had invested $1 at an interest rate of 3%, but that this interest was to be compounded m times a year, i.e., (3/m)% interest on the balance of the account would be awarded at the end of each of m equally spaced intervals during every July 4 to July 4 year. We want to make a table of the values of the account in the year 2000 as a function of m, where m varies from 1 to 12.

We have already computed the desired values for m=1, m=2, and m=3. Record these answers below.

When m=1, value of the account in 2000 is........
When m=2, value of the account in 2000 is........
When m=3, value of the account in 2000 is........

We will now modify the computations done in Sections 6 and 7 to compute the account values for the remaining values of m. That is, we will view m as an arbitrary positive integer equal to the number of times we compound in one year.

We again recall the meanings of k and A(k):

k = the number of interest periods which have gone by (on any given date);
A(k) = the amount of money in the account at the end of the k-th interest period.

First record the values for A(0), A(1), A(2), A(3),... . Place your answers below.

A(0) = ---
A(1) = ---
A(2) = ---
A(3) = ---

Do you see a general pattern emerging? Use this pattern to define A as a Maple function of k. Complete and Enter the following function definition statement.

```
> A :='A':  k := 'k':  m :='m':
> A := k -> ??? ;
>
```

Now determine k for the year 2000, i.e., the number of investment periods which will have occurred by July 4, 2000. Place your answer below.
k = ???

Use your formula to compute the value of the account in the year 2000. Enter your computations below.

Thus, for each m (number of compounding periods in one year), we can compute f(m), the final balance of our account in the year 2000. Write down the formula for f(m). Complete and enter the following function definition statement.

```
> f := 'f': m := 'm':
> f := m -> ??? ;
>
```

Now compute a list of values [m, f(m)] showing how the final balance in the account varies with the number of compounding periods. We place our values for m = 1,...,12 into a data list compoundData. Complete, then Enter, the following data list.

```
> compoundData := [ [ 1, ???], [ 2, ???], [ 3, ???], [ 4, ???], [ 5, ???], [ 6, ???],
>                   [ 7, ???], [ 8, ???],  [ 9, ???], [10, ???],  [11, ???], [12, ???]];
>
```

Maple Tip: You can perform separate computations for each m, then type the values into the list one at a time, that is the simplest approach, although a bit tedious, as you have seen above. A more sophisticated method is to have Maple construct the full data list with one command of the form:

compoundData := [seq([m,f(m)], m=1..12)];

Complete the following line and execute it.

```
> compoundData :=??? ;
>
```

(The seq command was first used in the section MORE ON MAPLE in Lab 1.)

Plot the data list compoundData. (If necessary, review the various plotting commands discussed in Part 9 of the MAPLE TUTORIAL in LAB 1 or Part 1 of LAB 2. Note: We are plotting a set of data points, so it would be natural to use the option "style=POINT", but also try it without that option.

Type and Enter your commands below.

```
> plot(???);
>
```

What seems to be happening as the compounding periods increase? Place your answer below.

Answer:

¤ 9. What is the best number of yearly compounding periods?

We would like to investigate compounding periods more frequent than m=12 times per year. As you saw in the previous section there is a simple formula for the balance f(m) of the bank account in the year 2000 as a function of the number m of yearly compounding periods. Type and Enter this as a Maple function again. Remember to Clear any values from the function name f before defining the function. Type and Enter your function below.

```
> f := 'f': m := 'm':
> f := m -> (???)^(224*m);
>
```

Use your formula to find the largest amount you could have received (rounding to the nearest cent) under any compounding scheme. Place your answer (and justification) below.

Maple Tip: In order to answer this question it will be necessary to generate more digits for f(m) than the Maple default of 10. This is done by the Digits command; for example, Digits := 18; will show 18 digits for the value of the quantity x.

```
> Digits := 18;
>
```

Answer to question above:

What is the minimum number m of compounding periods per year that would achieve this maximum balance? (Hint: If you claim the largest amount after rounding is $755.50, then you need to find the smallest number m which gives f(m) > 755.495 you do this with a "manual search" in that you compute f(m) for various m values until you find what you need.) Place your answer (and justification) below.
Answer:

What is the length of time of each compounding period to achieve this maximum return on the investment? Choose the most appropriate unit of time. Place your calculations and answer below.
Answer:

¤ 10. What if there is an additional charge?

It has been discovered that your ancestor included a service charge payment to the bank, payable at the end of the account on July 4, 2000. The one-time, lump-sum payment is to be: one cent for each time the bank computed and added an interest payment into your account.

Assuming m=1, how much money will you now receive on July 4, 2000? Enter your answer below.
Answer:

Assuming m=2, how much money will you now receive on July 4, 2000? Enter your answer below.
Answer:

Let g(m) denote the amount of money you will receive on July 4, 2000, assuming m annual compounding periods. Give a formula for g(m). Complete and Enter the following function definition statement.

```
> g := 'g':  m := 'm':
> g := m -> ??? ;
> # Scroll down with down arrow.
```

Given the service charge of one cent per compounding period, determine the value of m which will maximize your profits on July 4, 2000. How much money will you receive with this value of m? Place your computations and answer below.

¤ 11. But could it really happen?

Yes! It really has happened!

In 1790, Benjamin Franklin left 2,000 pounds (about $8,880) to be divided among Philadelphia, Boston, and their respective states. Part of the accumulated balance was to be paid out in 1890; the rest, in 1990.

The 1890 Boston disbursement was marked by a 14 year legal battle for control of the money.

In 1990, Boston and Massachusetts' share was estimated to be worth $4.5 million. Philadelphia's share was estimated to be worth $520,000. (Financial mismanagement, anyone?)

("Franklin's largesse has long reach," Boston Globe, April 17, 1990, pp. 15 and 17.)

In 1921, Maine's own James Phinney Baxter left Boston $50,000 in trust. His son Percival Baxter left $200,000 in trust in 1969. Boston was to start building a pantheon honoring New England settlers within one year of the trust reaching $1 million. If Boston failed to do so, then the money was to go to Portland.

In June 1992, the funds were worth $2 million, and Portland filed suit against Boston for the money, claiming that the funds had passed the $1 million mark more than a year previously.

("Portland wants Boston to forfeit $1 million," Portland Press Herald, January 16, 1992, pp. 1A and 12A.)

("Portland will take Boston to court over trust funds," Portland Press Herald, June 4, 1992, p. 1B.)

LAB 3B: Loans: Interest, Principal & Payments

□ NAMES:

□ PURPOSE.
In this lab we explore questions related to the calculation of interest payments for a long-term loan. We have three goals for this lab:
(1) To learn to use recursive definitions;
(2) to explore a particular type of limiting behavior; and
(3) to learn additional features of Maple.

□ PREPARATION.
Read Chapter 1 of Calculus: Modeling and Application. This provides general information concerning functions and relationships between two variables. Of particular relevance to this lab is the material in the first few pages of Section 1.8 discussing car payments.

Enter the following line to initialize Lab 3B..

```
> with(lab3b);
> # If an Error is an output of the above line, see your lab instructor.
```

□ LABORATORY REPORT INSTRUCTIONS.
Follow you instructor's instructions if they differ from ours.

For this project you are not required to submit a formal report. Rather you are asked to submit an electronic copy of this Worksheet with the questions answered in the indicated locations. You will be graded on this single submission. Please write in complete sentences.

□ PROJECT.

¤ 1. A Practical Story of Interest.
Suppose you want to buy a house at the price of $75,000. There is a 10% down payment of the sale price; you will pay the rest off over 30 years at an annual interest rate of 8%. You want to determine the monthly payment.

What is your initial guess as to what your monthly payment will be?

Place your guess here: ____________________

¤ 2. What is the monthly payment?

Let's assign symbols for the important variables in this problem.

p = the monthly payment, the quantity we wish to determine.

$P(k)$ = the "principal" after k months, i.e., the dollar amount you still owe after making payments for k months, where k varies from 0 to $n = 12*30 = 360$, the total number of monthly payments to be made.

A few more facts are needed. Each monthly payment p is a sum of two parts:

$$p = I(k) + R(k)$$

where I(k) is the interest your loan generated during the k-th month, and R(k) is the portion of the payment left over to reduce the principal.

To make this less abstract, let's proceed month by month. What is the starting principal P(0), i.e., the amount that you initially have to borrow from the bank? Place your answer below.

P(0) = ---

At the end of the end of the month (i.e., k=1), how much interest will be owed on this loan (remember, the annual interest rate is .08). Place your answer below.

I(1) = ---

Hence, at the end of one month, how much of the payment p will go towards reduction of the principal? (Note: your answer must involve the unknown p.) Place your answer below, and then check it by Entering the following line.

R(1) = ---

```
> Ans(1);
>
```

Give a formula for the new principal P(1) after the first monthly payment. Place your answer below.

P(1) = ---

So we have succeeded in going from P(0) to P(1) (although we must tolerate the presence of the unknown p). Now let's go from P(1) to P(2). At the end of the second month (i.e., k=2), how much additional interest will be owed on this loan? Place your answer below.

I(2) = ---

Hence, at the end of the second month, how much of the second payment p will go towards reduction of the principal? Place your answer below.

R(2) = ---

Give a formula for the principal P(2) after the second monthly payment. Place your answer below.

P(2) = ---

These formulas should be rapidly becoming unpleasant! However, the general method for computing the loan principal should be emerging from our calculations. Let's apply this method to see how to proceed from the principal P(k-1) to the principal P(k). During the k-th month, how much additional interest will be owed on this loan? Place your answer below.

I(k) = ---

Hence, at the end of the k-th month, how much of the k-th payment p will go towards reduction of the principal? Place your answer below.

R(k) = ---

Give a formula for the principal P(k) after the k-th monthly payment. Place your answer below.

P(k) = --- ??? , in terms of P(k-1) and R(k),
P(k) = --- ??? , in terms of P(k-1) and p.

Warning!! Write P(k-1) ONLY ONCE in your recursion equation for P(k)!! More than one copy will slow down the computation of P(k). Why? Well, suppose you have two copies of P(k-1) in your recursion formula. Then executing P(60) will call for the value of P(59) twice. But each call for P(59) will require the value of P(58) twice (for a total of four calls). In general, there will be 2^k calls for the value of P(60-k). This quickly becomes astronomical.

Check your answer for P(k) by Entering the next line.

```
> Ans(2);
>
```

You have obtained a recursion formula for P(k), expressing this quantity in terms of P(k-1). We now use Maple to evaluate P(k) for certain values of k, including k = 360 (30 years). First P(0) (Don't forget the down payment!). Complete and activate the next command.

```
> P(0) := ??? :
>
```

You can now place your recursive formula into a Maple "do" command and compute P(n), for n a positive integer.

Maple Tip. Be sure to write the numbers in your formula for P(k) with decimal points, i.e., ".08" not "8/100", the decimal point tells Maple to use numerical approximations. Without the decimal point, many computations in this project will not be as quick.

```
> p := 400 :   n := 4 :
> for k from 1 to n do
>    P(k) := P(k-1)*???: od:
> # The od is Maple code for finishing the do command.
> P(n);                          #This will print out just the last value, P(n).
>
```

Now, given a value for the monthly payment p, you could replace p in the recursively defined function P and compute the principal remaining at the end of the loan period, P(360). Our goal, however, is an inverse problem: given the value P(360)= 0 (i.e., loan is repaid in 30 years), determine the value of the monthly payment p. Determine this desired value of p to the nearest cent by taking guesses for p and computing the resulting P(360). To start, try a specific payment like p= 400 and a small value for n, like n=4 to see if everything is working all right.

Record your answer for p below along with supporting computations.

p = --- .

¤ 3. How much interest will you pay?

Over the 30 year lifetime of the loan, how much will you pay in interest? (Hint: All the money you pay is either interest or principal, and the total principal is merely the original principal of the loan.) Place your calculations and answers below.

¤ 4. Is there an explicit formula for the monthly payment?

Yes! The recursive formula that you developed in Section 2 can be replaced by an explicit formula, i.e., a formula that does not need to be applied multiple times in order to produce an answer. (The value of such a replacement in terms of speed-of-computation is great.) The explicit formula we desire is the first formula in Section 1.8 of Calculus: Modeling and Application. (The derivation of this formula is not given here, it requires techniques from second semester calculus.)

The formula from your text is

```
        (P - D)*r*(1+r)^n
p = ------------------------ ,
          (1+r)^n - 1
```

where

p is the monthly payment,
P is the original price,
D is the down payment,
r is the monthly interest rate (i.e. .08/12 for 8 % annual rate), and
n is the number of months required to pay back the loan.

Use this formula from Calculus: Modeling and Application to calculate our monthly payment p. Check your formula by using it to calculate p and comparing the answer to that obtained recursively in Section 2. (If the two answers don't agree, then check your work for an error.) Enter your formula and computations below.

¤ 5. Can you afford a more expensive house?

The formula referred to in Calculus: Modeling and Application allows for the variation of four quantities in the computation of monthly payments p. Let's incorporate some of this flexibility into our formula in order to examine other house loan options.

Suppose we want to look at a variety of houses: in that case, the price of the house will vary. We'll denote this variable by P.

We'll continue to assume that a 10% down payment is required (although we could easily vary this as well), but we wish to allow for variation in the interest rate. We'll denote the yearly interest rate by R. (The monthly interest rate is thus R/12.)

Finally, we might want the loan period to be shorter than 30 years (e.g., we expect our daughter to enter Bowdoin in 15 years--when the tuition will be $150,000 per year--and we don't wish to be saddled with mortgage payments at the same time). We'll let "term" denote the length of the loan period in years. (Thus, the length of the loan period in months is 12*term.)

Maple Tip: It is often convenient to stringing Maple commands together on one line, separating the different commands by colons or semi-colons. Upon Entering such a string, all the commands will be executed, but the output will be shown only for those followed by a semi-colon. This is particularly useful for specifying values for variables, followed by a quantity that you wish to evaluate. An example would be

```
> a :=1:   b :=2:  c :=3:    d := a+b+c;
>
```

Determine a Maple expression for the monthly payment p in terms of P, R, and term. Type and Enter your formula below.

```
> P := ???:  R := ??? :  term := ???:
> # A convenient method of changing the input.
> r := ???:   n := ???:              # This keeps the next formula a bit friendlier.
> p := ???;                           #A Maple expression, not a function.
>
```

Check your formula by using it to compute the monthly payment we determined earlier for a $75,000 house at 8% interest over a 30 year period. You do this simply by giving Maple the values for P, R, and term, and then asking for the resulting value of p (i.e., just type and Enter the one letter p--Maple will return the desired number). Enter your calculations below.

```
> P := ???   : R := ??? :   term := ???   :
> r := R/12:  n := 12*term:
> p := ??? ;                        # Use Copy and paste to get this.
>
```

Suppose you find a $150,000 house that you like better than the $75,000 house. If all the other loan parameters remain the same, how much will your monthly payments be? First make an educated guess as to what the payment will be, then compute it exactly. Any surprises?

Place your conjecture here: ____________________ Place your computations and answer below.

```
> P := ???  : R := ??? :  term := 30  :
> r := ???: n := ???:
> p := ??? ;
>
```

¤ 6. How serious is an interest rate increase?

Suppose, while you are trying to decide about buying the $75,000 house (not the $150,000 house, which you've determined you can't afford), the interest rate increases to 9%. Conjecture what the new monthly payment will be, and then check your conjecture with the formula.

Place your conjecture here: ____________________ Enter your computations and answer below.

¤ 7. What is the nature of the dependence on interest rate?

How did your conjecture and your calculation in Section 6 compare? Is the monthly payment at 9% interest merely 9/8 times the monthly payment at 8% interest? If not, explain why the payment at 9% is less than (or greater than) 9/8 times the payment at 8%. Place your answers below.

"The payment at 9% interest is": ---

"9/8 times the payment at 8% interest is": ---

Plot the monthly payment p as a function of the interest rate R (.001<=R<=.30), holding P=75000 and term=30. You can do this most easily by using a function payment(R) defined by the formula

```
> R := 'R';
> payment := R -> p;
> plot(payment, ???);
>
```

Be sure, however, that you first define P and term to be the appropriate values. Any surprises in the graph? How do you explain the behavior? Enter your plot and answers below.

¤ 8. How much will an increase in interest rate cost you?

Over the 30 year life of the loan, how much will the delay in making a decision (during which time the interest rate rose from 8% to 9%) cost you? Place your calculations below.

¤ 9. (Optional) Other variations.

How does the monthly payment vary with the term of the loan, i.e., the number of years for repayment? Can you modify the formula for the payment to allow for a variable down payment percentage? In Section 7 we plotted and studied the payment p as a function of the interest rate R--repeat this process, replacing the interest rate R with the term of the loan. Any surprises here? Are there any other problems that interest you? Try to solve them. Enter your computations and answers below.

LAB 4: Limited Population Growth

□ NAMES:

□ PURPOSE.
In this lab we study the problem of modeling the population as a function of time in a setting where there is a maximum population M that the environment will support. We will start with an initial population value and an assumption about the nature of the growth rate (i.e., that the rate of growth slows down in a particular way as the population approaches its maximum level) and then apply a simple graphical method to predict subsequent population values. These population values will be predicted at a sequence of times Dt, 2Dt, 3Dt, ... , where Dt is a fixed time step. One goal of the lab is to learn how the population predictions change as the time step Dt is decreased.

¤ Comparison with the Rate of Growth Project.

In a recent classroom project you may have studied the growth of a fruit fly population, starting with data on the population over a series of days, and ending with a function that modeled the population as a function of time. Recall the methods and results of that project as you work through this lab. In particular, recognize the following differences between the project and this lab:

- The project data represented a situation in which there was no limit placed on the population size. In this lab we model a population where there are limits to the population growth rate.

- The project started from population data and ended with an equation modeling the population as a function of time. In this lab we start with a theoretical assumption about the nature of the population growth rate, and end with predictions (data) for future populations.

- In the project the time step was fixed at one day. In this lab we will study what happens to our predictions as the time step decreases.

□ PREPARATION.

- Study the Introduction to Chapter 2 and Section 2.5 of Calculus: Modeling and Application.

- Complete the Preliminaries section of the laboratory before coming to the lab session. Write down (on the printed copy of your lab) the answers to the questions. You will be able to check these answers during the lab session.

- Review your classroom project on the growth of fruit flies.

□ MAPLE Procedures.
It is necessary to activate the following line before continuing with the lab.

```
> with(lab4);
> # If an Error is an output of the above line, see your lab instructor.
```

□ LABORATORY REPORT: General Instructions.
Note: Follow your instructor's instructions if they differ from those given here.

You are to write a self-contained, well-organized exposition and analysis of the issues raised in this laboratory. In particular, it should incorporate the answers to all the important questions raised during the lab session.

Two submissions will be made for this report: an initial draft, and a final revision. Comments made on the draft will help you prepare the revision, and only the revision will receive a grade. Moreover, the grade will be based solely on the written report, the instructor will look at other parts of the worksheet only in unusual situations. You should therefore consider the entries you make in the Project section to be notes to help when later writing your report.

More specific instructions on the content of the report will be given at the end of the worksheet.

□ PRELIMINARIES.
¤ 1. Theoretical Preliminaries.

• The Theoretical Assumption.
When there is a maximum population M that the environment will support, biologists often assume that the rate of change of the population is proportional to the product of (1) the population and (2) M minus the population.

Write out this Theoretical Assumption using c to denote the proportionality constant, p to denote the population, and Deltap to denote the change in population in a time interval of length Deltat. Type your equation after "Answer," then check your answer by activating the line below it.

• Learning Tip.
For your effective learning it is important that you seriously attempt to answer questions before checking the answer.
Answer:

```
> Ans(1);
```

(We prefer to use the capital Greek letter Delta in place of the Delta in Deltap and Deltat but with this version of Maple that is not possible in text mode. Activate the next line to see the Greek letter.)

```
> Delta;
> # Scroll down with a down arrow.
```

Here are two reasons to justify assuming such a relationship:
(1) When the population p is small, M minus p is essentially the same as M (a constant), and then Deltap/Deltat is (essentially) proportional to just the population p; this is the "natural growth" situation considered in the text. Hence, for small populations, our new model should result in population predictions that will look a lot like "natural population growth." This behavior is what a biologist would expect to happen with most populations.

(2) At the other extreme, when the population gets close to M, the factor M minus p is close to zero, and this effectively shuts down further growth. Thus, the factor M minus p captures in a natural way the effect of a limited environment.

• The Goals of this Laboratory Exercise. Given values for the initial population p(0), the maximum population M, the time step Deltat, and the proportionality constant c, we wish to implement a method for predicting the population at times

t(0)=0, t(1)=Deltat, t(2)=2*Deltat, t(3)=3*Deltat,... .

Our predicted population values at these times will be denoted by

p(0), p(1), p(2), p(3), ...

These predictions will be based on the Theoretical Assumption stated above, we will then examine these predictions to see how "realistic" they are.

¤ 2. The Recursive Procedure.

Our method for estimating the population values will be recursive, i.e., p(k) will be determined by p(k-1) for each k=1, 2, 3, ... To begin the recursion, given p(0) (which is just the initial population), how do we obtain p(1)? Let's take some time to answer this.

The derivation is most easily understood by thinking geometrically in terms of a graph of population verses time. The following diagram shows a graph of the starting situation. Plotting population as the dependent variable and time as the independent variable, the initial population p(0) is represented by a vertical line segment of length p(0) at the starting time t(0)=0.

Let Deltap/Deltat be the rate of change of the population during the time period t(0)=0 to t(1)=Deltat. Thinking geometrically, Deltap/Deltat is the slope of the line segment joining the initial population point (t(0), p(0)) to the next population point (t(1), p(1)). Let's label this slope as slope(0).

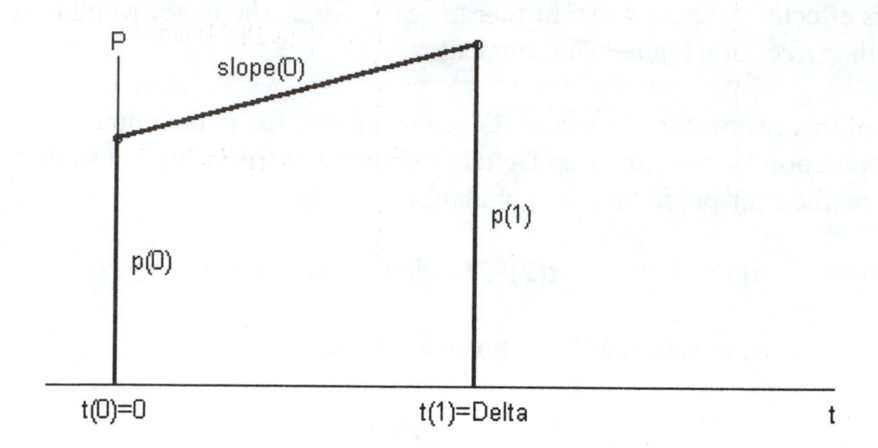

The slope of a line segment is the rise over the run. In our situation, the run is Delta = Deltat ...

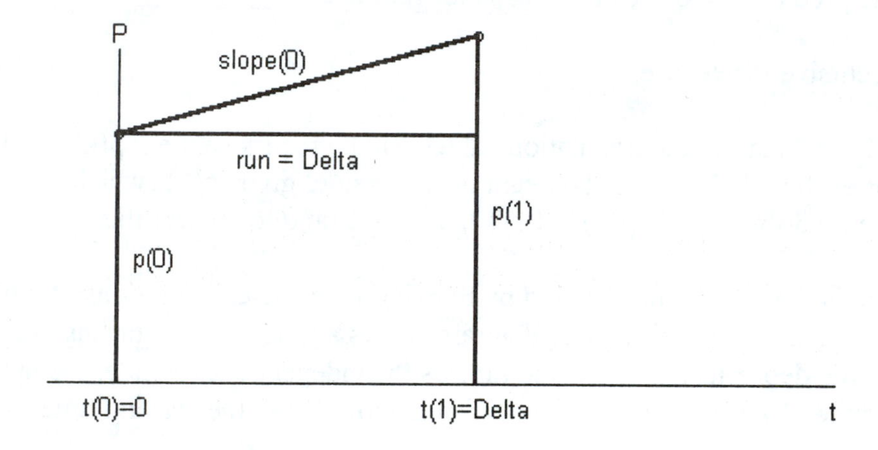

... while the rise is slope(0)*Delta.

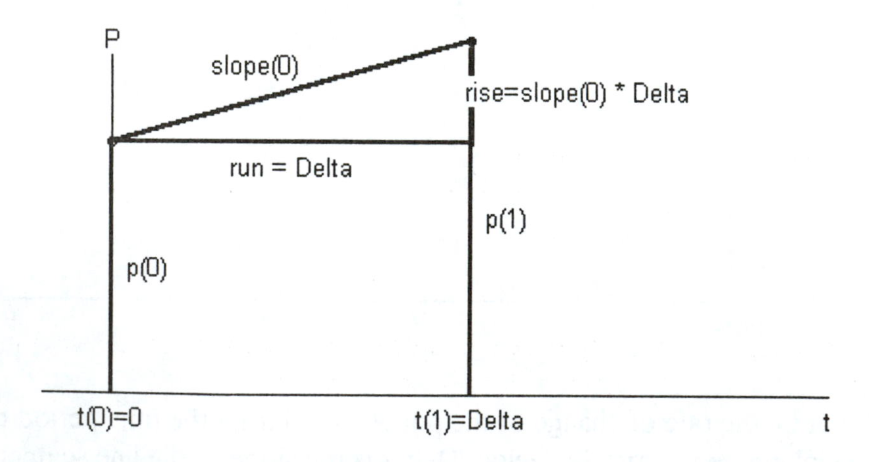

The line segment representing p(1) is the sum of two pieces, one of length p(0), and the other of length rise=slope(0)*Delta. Hence p(1)=p(0)+slope(0)*Delta, as shown below.

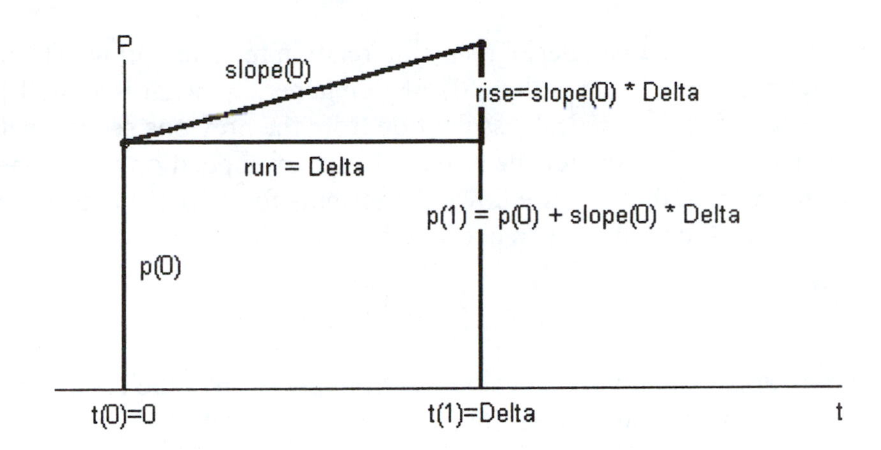

Thus we have a formula to compute p(1) from p(0). Give the resulting equation.

Write your equation for p(1) on paper, then check it.

```
> Ans(2);
> # Scroll down with the down arrow.
```

• Maple Tip. The above pictures have been put into an animation (in the hopes that it will make the preceding derivation more understandable). To see the animation, activate the following Maple command after reading the following discussion. The lower control panel in the animation plot starts with a square (stop), followed by a right pointing triangle (run animation), followed by an arrow pointing into a vertical line (one picture at a time). You can restart the animation by clicking on the 1:1 button. The next two commands either slow down or speed up the animation, and the last two on that line can be toggled back and forth; experiment with them.

Use the ?animation command to learn how to display the individual plots that are shown above. For example animation(construct=1) will display the first plot.

```
> animation(now);
>
```

The same reasoning applies to computing p(2) from p(1). Give the resulting equation. Write your equation for p(2) on paper, then check it.

```
> Ans(3);
```

The same reasoning applies to computing p(k) from p(k-1) for any k between 1 and n. Give the resulting equation. Write your equation for p(k) on paper, then check it.

```
> Ans(4);
```

Great! It looks like we have a wonderful recursion relation for computing p(k) from p(k-1), except for one little problem: What is slope(0)? More generally, what is slope(k) for k=1,2,... ? Answer: That is where our Theoretical Assumption from the previous section enters the picture! The assumption gives us a formula for the rate of change of population (the slope!) in terms of the current population. Use this to write down the formula for slope(0) in terms of p(0). Write your equation for slope(0) on paper, then check it.

```
> Ans(5);
```

The same reasoning applies to computing slope(k) from p(k). Give the resulting equation. Write your equation for slope(k) on paper, then check it.

```
> Ans(6);
>
```

We thus have an interesting interweaving, we compute our quantities in the following order:

p(0), slope(0), p(1), slope(1), p(2), slope(2), p(3),...

Summary for latter convenience. (complete this)

M := ???; c := ??? ; Delta := ???, p(0) := ??? ;
slope(0) := ???
p(k) := ???
slope(k) := ???

• Maple Tip: Have you saved your work recently? If not, do so now!

□ PROJECT.

¤ 1. A Computational Example.

As in the classroom project we'll assume that we have a population of fruit flies which initially equals 111 flies. However, we will now add the new assumption that the maximum population supported by the laboratory environment is 1000. In order to carry out our computations we will also need a value for the proportionality constant c: the value c = .000098 best matches the data from the classroom project (you will show this at the end of the lab). We will initially start with population measurements taken every 5 days, i.e., Delta = 5.

Fill in the missing values for the following table using the formulas from the Preliminaries entered as Maple commands below the table. (The final row of values is supplied in order for you to be able to check your calculations.)

Time	Population	Slope = Dp/Dt
t(0)=0	p(0)=111	slope(0)=
t(1)=	p(1)=	slope(1)=
t(2)=	p(2)=	slope(2)=
t(3)=	p(3)=310.435	slope(3)=20.9784

```
> M :=1000:  c:=.000098:  Delta :=5:
> p(0) := 111: slope(0) := c*p(0)*(M-p(0));
> p(1) := p(0) + slope(0)*Delta;    # Change indices as needed.
> slope(1) := c*p(1)*(M-p(1));
>
> # Scroll down with down arrow.
```

• Maple Tip.
It may seem strange to report non-integer population values (How can we have .435 of a fly?). However, rounding these numbers during the computations can have undesired consequences, so we retain the decimal values. Rounding to integer values can always be done later.

In what follows we will use Maple to "automate" this process and to examine what happens when the value of the time interval Delta is changed.

¤ 2. A Computational Example with Maple.

• A. Input the maximum population M=1000, the initial population p(0)=111, and the proportionality constant c=.000098.

Complete and Enter the commands on the line below.

```
> M :=   ??? ;        p(0):= ??? ;        c :=   ??? ;
>
```

• B. We wish to examine the time period from 0 to 100 days using a variety of time intervals Delta. To do so we will simply specify n, the number of (equal) time intervals that we wish to create from (0, 100) . Then Delta := 100./n. Input the (initial) value n=20 and the equation for Delta as a function of n.
Complete and Enter the commands on the line below.

```
> n :=   ??? ;        Delta := ??? ;
>
```

• C. We now define the time values, t(0), t(1), ... , t(n). As a preliminary task, fill in the appropriate values in the following list (remember that the distance between consecutive t values is delta):

t(0) = ___
t(1) = ___
t(2) = ___
t(3) = ___

Given the entries in your list, do you see what the general formula should be, i.e., the formula for t(k) in terms of k and delta? Type it here:

t(k) = __________ for k = 0, 1, ... , n.

We need to write this in Maple functional notation. Complete this on the next Maple command line, enter it, then check it by activating the Ans(7) command.

```
> t := k -> ???;
> Ans(7);
```

Complete and Enter the command on the line below.

```
> seq(t(k), ???);
>
```

You should obtain with the last command a list of integers from 0 to 100 by 5's. Try again if you did not.

• D. To begin the recursion we must Enter the initial value of the slope, slope(0). You worked out the formula for slope(0) at the end of the Preliminaries.
Complete and Enter the defining command for slope(0) on the line below.

```
> slope(0) := ???
> # Scroll down with the down arrow.
```

• E. We need to define the subsequent slope values,

slope(1), slope(2), ..., slope(n),

as well as the population values

p(1), p(2), ..., p(n).

Since these are recursively defined, there are advantages in using the Maple method of defining functions with the procedure, "proc" command. This is the main way in the Maple programming language for creating a computer program or procedure.
Activate the following commands:

```
> slope :=proc (k)
>       slope(k) := c*p(k)*(M-p(k));
>       end:
> p :=proc (k)                          # Defines population at time k, p(k).
>       p(k) := p(k-1) + slope(k-1)*Delta;  # Note the p(k) is defined in terms of
>       end:                            # p(k-1) and slope(k-1). Therefore,
> p(0):=111;                            # to get started we define p(0) and slope(0).
> slope(0) := c*p(0)*(M-p(0));
>
```

We can use a "seq" statement again, but we must treat both slope(k) and p(k) in the same seq statement, not in separate ones. Why? Because as seen earlier, slope(k) and p(k) are interlaced: p(0) is needed for slope(0), which is needed for p(1), which in turn is needed for slope(1), etc.

Fortunately the seq command allows for pairs of statements by use of a list: [expr1, expr2] is handled by evaluating expr1 first and then expr2 second.

You derived equations for p(k) and slope(k) at the end of the Preliminaries; place these equations in the following "seq" command. The iterate k will eventually run from 1 to n.

Warning!!! It is critically important to test a recursion relationship before running it too many times. For that reason you will first test your seq command with k running from 1 to 4.

Complete and Enter the "test" seq command on the line below.

```
> seq([p(k), ???], k=1..4);
>
```

Important!!! Your result should be a list of four pairs of numbers. If anything else appears, then do not execute the "seq" command for k larger than 4 until you fix the problem!!

If you have a problem... Note the names of any non-numerical symbol which appears in your output (e.g., p(0), c*p(0), M, etc.). One of the following problems has probably occurred:

(1) the symbol should have been assigned a value in the previous steps (which for some reason did not happen) or

(2) the symbol was incorrectly introduced by an error in the statements you typed. These two problems most often occur for the following reasons:

(1) You did not Enter the statement which assigned a value to your symbol. If so, then go back, find the necessary statement, and Enter it.

(2) You may have forgotten that multiplication is designated with a * .

For this reason c*p(k) denotes the multiplication of c and p(k). But cp(k) does not indicate a multiplication--it indicates a new indexed variable cp. Correct any errors of this nature in your seq commands. Continue to correct and reenter your seq (with k going up only to 4) until the resulting output is a list of four pairs of numbers. Correct and reEnter your seq command above as needed.

If you do not have a problem... You are now ready to run your seq command for the full range of k values. Simply change k=1..4 to k=1..n, and then Enter the command.
Enter your seq command with the full k range.

```
> seq([p(k), ???], ???);
>
```

This results of this computation will be held in your computer's memory long enough for us to list and then graph the time, t(k), verses population, p(k) data and then the time verses slope data.

• F. We now create a list of data points [t,p] for population p(k) as a function of time t(k) by using the seq command. We call this data list popData(n). Note, in particular, the values of the population at time t=20 days and t=100 days.

Enter the command shown on the line below.

```
> popData(n) := [seq([t(k),p(k)], k=0..n)];
>
```

• G. Create a scatter plot of data points for population p(k) as a function of time t(k) by using the plot command. Input your scatter plot here; call it popGraph(n).
Complete and Enter your plotting command on the line below.

```
> popGraph(n) := plot (???): ";
>
```

Do the population values vary in the way you expected?

Answer:

• H. Create a list of data points for slope (rate of change) slope(k) as a function of time t(k) by using the seq command. Call this data list slopeData(n).
Enter your command on the line below.

```
> slopeData(n) = [seq([t(k), ???], k=???];
>
```

• I. Create a scatter plot of data points for slope (rate of change) slope(k) as a function of time t(k) by using the plot command. Input your scatter plot here; call it slopeGraph(n).
Enter your plotting command on the line below.

```
> slopeGraph(n) := plot(???): ";
>
```

Do the slope values vary in the way you expected?
Answer:

¤ 3. Decreasing the Time Intervals

Redo all of Section 2, changing the number n of time intervals to 40, then 80, then 160, and finally to 320.

No, wait! Don't drop the course! This is easy! We have already copied all the relevant Maple commands from Section 2 and placed them together in three
regions below.

• Maple Tip. You can activate a whole region at once by placing the cursor anywhere in the region and hitting Enter.

```
> M := 1000:    c := .000098:
> n :=320:   Delta := 100./n:
> t := k -> k*Delta:
> slope :=proc (k)   slope(k) := c*p(k)*(M-p(k));   end:
> p :=proc (k)   p(k) := p(k-1) + slope(k-1)*Delta; end:
> p(0):=111:
> seq( [p(k), slope(k)], k=1..n):
> popData(n) := [seq([t(k),p(k)], k=0..n)]:
> popGraph(n) :=plot(popData(n),style=POINT,symbol=CIRCLE,color=blue): '';
> slopeData(n) := [seq([t(k),slope(k)], k=0..n)]:
> slopeGraph(n) := plot(slopeData(n), style=POINT, color=blue): '';
>
```

Change the one statement in your gathered collection that defines the value of n.
Activate the region above that contains all the gathered commands, then watch Maple compute.
Repeat this process for all the n values 40, 80, 160, and 320. Activate the gathered commands cell for each of the n values.

The next command produces a list of your predicted population sizes at t=20 days for each of the five n values analyzed above (n = 20, 40, 80, 160, and 320). How does the increasing of n (hence decreasing of the time step size Delta) affect the predicted population sizes?

Read the following Note then Enter the command.

Note: The following command will work only if you have run your full set of gathered commands for all the values n = 20 (done in previous section), 40, 80, 160, and 320.

```
> seq([n, popData(n)[1+n/5][2]], n=[20,40,80,160,320]);
>
```

Question: Explain why the notation popData(n)[1+n/5][2] represents the predicted population on day 20 for each of the values of n.

Answer.

The following command produces a table of your predicted population sizes at t=100 days for each of the five n values analyzed above (n = 20, 40, 80, 160, and 320).

```
> seq([n, popData(n)[n+1][2]], n=[20,40,80,160,320]);
>
```

How does the increasing of n (hence decreasing of the time step size Delta) affect the predicted population sizes? Place your answers below.

Answer:

¤ 4. Analysis.

• A. What happens to the models as the number of time steps n increases (i.e., as the time step Delta decreases)? Which model do you think most accurately portrays the real situation? Why? Place your answers below.
Answer:

• Maple Tips.
1. To help in making comparisons, place the n=20 population plot and the n=320 plot into the same picture via the plots[display] command. This command will be of the form plots[display]([nameOfPlot1, nameOfPlot2]);
2. Both of these need to be graphed first before placing them in the next command line. To better see the comparison, you might first re-graph the n=20 case where you add the option "symbol=CIRCLE" after the "style=POINT" option. You could also change the color to blue. Then you could see the one graph in blue and the other in red.

```
> plots[display]( [popGraph(20), popGraph(320)] );
>
```

• B. Estimate the time at which the fly population is increasing most rapidly, and estimate the size of the fly population at this time. (Hint: Look at your slope graphs!)
Place your answers below.

• C. In the classroom project on the growth rate of fruit flies we saw that, in an unconstrained environment, the fruit fly population satisfied a difference equation of the form

Deltap/Deltat = k*p

where k was approximately equal to .098. However, our current difference equation model is

Deltap/Deltat = c*p*(M-p).

Compare these two difference equations when p is much smaller than M (the maximum sustainable population), using the comparison to show why we chose the value of the proportionality constant c to be .000098.
Place your answers below.

□ LABORATORY REPORT-Specific Instructions.
Note: Follow your instructor's instructions if they differ from those given here.

Write an exposition of the lab experience. Start by explaining in your own words how we get from "slope = rise/run" to the recursive formulas that allow us to step from one population value to the next. This is a method of critical importance in this course, so you need to fully understand it. You don't need as much detail as we provided in our Preliminaries discussion; one or two paragraphs should suffice.

Then describe your experience with the computational part of the lab. Your exposition should include answers to the questions we raised, but it should not be just a list of answers. In particular, your report should have a unifying theme (e.g., "The essential goal of this lab is to... .") that ties together the findings of the lab procedures.

The clarity of your explanations is at least as important as your answers themselves--clear and unambiguous statements of goals, procedures, and conclusions are what we will look for. Our working definition of good expository writing is simple: the writing is good if we easily understand your intended meaning, the writing is weak if your intended meaning is unclear. Furthermore, your report should make sense to a reader who does not have the rest of this lab worksheet and who was not present with you in the lab--that is, the report should be self-contained and independent of the rest of the worksheet.

There is no required length for the report. However, in order to include and adequately discuss all the relevant items, your report will probably be at least the equivalent of one or two printed pages. You should critically examine your work if the report turns out to be less than one page.

• Maple Tip: One good way to proceed is to print a copy of your completed worksheet for reference as you write the report (otherwise you might have to scroll back and forth endlessly in the worksheet). However, if you choose to print, please follow the printing instructions that were distributed at the beginning of the semester. Alternately you could copy your completed Maple file, say Lab4.ms to another file, say Ans4.ms . Then it is an easy matter to delete all the unnecessary material from this second file before enhancing it for your final report.

LAB 5A: Rates of Change--Puppy Growth

□ NAMES:

□ PURPOSE.
The primary purpose of the laboratory will be to learn how to detect certain types of growth patterns in a given data set. The primary tools will be semilog and loglog plots. These tools will be applied to data on the growth of a puppy, and the results will be used to construct a formula to model the observed pattern in the puppy's growth.

□ PREPARATION.
To fully understand the theory employed in this lab you should study Section 2.6, "Logarithms and Representation of Data," in Calculus: Modeling and Application. You should also review the Maple procedures that were discussed in Lab 1.

□ MAPLE Procedures. Please Enter the following command.

```
> with(lab5);
> # If an Error is an output of the above line, see your lab instructor.
```

There are two procedures written for this lab. They are DoubleListPlot and DataExample. For more information on these use the ?DoubleListPlot and ?DataExample commands at the Maple command line.

□ LABORATORY REPORT INSTRUCTIONS.
Note: Follow your instructor's instructions if they differ from those given here.

For this project you are not required to submit a formal report. Rather you are asked to submit a copy of this Worksheet with the questions answered in the indicated locations. You will be graded on this single submission. Write in complete sentences and be sure the sentences connect together in coherent ways.

□ PRELIMINARIES.
¤ 1. Data sets exhibiting exponential relationships.

A data set [[t1,y1], [t2,y2], ..., [tn,yn]] is said to exhibit an exponential relationship if there is an exponential function y = A*e^(k*t) such that yi = A*e^(k*ti) for every data point [ti,yi]. This is equivalent to observing that all the data points [ti,yi] lie on the exponential curve y = A*e^(k*t). Enter the following command (a few times!) for examples.

```
> DataExample(exp,exact):
>
```

Rarely will real data, i.e., data measured from a real world phenomena, actually fall precisely on an exponential curve. The best we can hope for is that the data will be well-approximated by an exponential curve y = A*e^(k*t). The formula

y = A*e^(k*t)

is then said to be an exponential model for the relationship between the variables t and y. Enter the following command (a few times!) for examples.

```
> DataExample(exp,approx):
>
```

¤ 2. Data sets exhibiting power relationships.

A data set [[t1,y1], [t2,y2], ..., [tn,yn]] is said to exhibit a power relationship if there is a power function y = A*t^r such that yi = A*ti^r for every data point [ti,yi]. This is equivalent to observing that all the data points [ti,yi] lie on the power curve y = A*t^r. Enter the following command (a few times!) for examples.

```
> DataExample(power,exact):
>
```

Rarely will real data, i.e., data measured from a real world phenomena, actually fall precisely on a power curve. The best we can hope for is that the data will be well-approximated by a power curve y = A*t^r. The formula y = A*t^r is then said to be a power model for the relationship between the variables t and y. Enter the following command (a few times!) for examples.

```
> DataExample(power,approx):
>
```

¤ 3. More general data sets.

We will first consider some data sets which are not well-approximated by either an exponential nor a power relationship. In such cases we can still try to fit exponential and power curves to the data, but the results are generally (but not always) bad. (Sometimes the "badness" of a fit will be overlooked at first, especially when a model curve seems to be "close" to all the data points. "Badness" might then be an obvious "trend" in the data points which is not matched by the model curve, e.g., the data points seems to be leveling off while the model curve predicts continued growth. Look for such contradictory trends when analyzing the examples.) Enter the following commands (a few times!) for examples.

```
> DataExample(norelation,approx):
>
```

Optional. Some of the examples you just generated will be pretty extreme: the attempted exponential and power function fits will be obviously terrible. However, some of the examples will be more subtle. Generate one of these "more subtle" examples and describe why you think each of the two proposed model curves are either good or bad. (Reread the parenthetical comment on "badness" in the previous paragraph.) Place your answers below.
Answer:

A reminder. Have you Saved your file yet? If not, do so now!

□ PROJECT.
¤ 1. Semilog Plots.

The output of the command DataExample consists of two lists tdata and ydata (the t coordinates and the y coordinates of the data points). These can be recombined in various ways by another new command: DoubleListPlot. Enter the following example.

```
> data := DataExample(exp,exact): tdata := op(1,data): ydata := op(2,data):
> DoubleListPlot(tdata, ydata);
>
```

The second plot is the output of the DoubleListPlot command. It is merely a scatter plot of the data set gotten by combining the t values of tdata with the corresponding y values from ydata (which is, in this simple case, the scatter plot of the original data set).

The primary value of DoubleListPlot is that it makes easy work of constructing semilog and loglog plots for a data set. We consider semilog plots now, and loglog plots later.

Recall that a semilog plot for a [t,y]-data set is merely a plot of the points [t, log(y)], i.e., take the logarithm of the y coordinate. Thus, to use DoubleListPlot to construct a semilog plot, we have only to take the logarithm of all the values in ydata. Enter the following example of a semilog plot.

• Maple Tip. In Maple the "ordinary" logarithm function, i.e., with base 10, is denoted by log[10](x), while the "natural" logarithm, i.e., with base e, is denoted by log(x). We will use the "natural" logarithm in what follows, although the process we now describe would work equally well with any choice of the base for the log function.

```
> data := DataExample(exp,exact):   tdata := op(1,data):   ydata := op(2,data):
> DoubleListPlot(tdata, map(log, ydata), title = `SemiLog`);
>
```

What is the most striking feature of the semilog plot that you have just generated? Enter the exponential example commands a few more times--do the subsequent semilog plots continue to exhibit the "striking feature"? Place your answer below.
Answer:

What does a semilog plot look like for data which is only approximated by an exponential relationship? Let's examine some examples. Enter the following commands (a few times!) for examples.

```
> data := DataExample(exp,approx):  tdata := op(1,data):  ydata := op(2,data):
> DoubleListPlot(tdata, map(log, ydata), title = `SemiLog`);
>
```

How would you describe the semilog plots that you just generated? Said another way, if a data set is well-approximated by an exponential curve, then what do you expect to find in its semilog plot? Place your answer below.
Answer:

¤ 2. Loglog Plots.

Recall that a loglog plot for a [t,y]-data set is merely a plot of the points [log(t),log(y)], i.e., take the logarithm of both the t and y coordinates. Thus, to use DoubleListPlot to construct a loglog plot, we have only to take the logarithm of all the values in both tdata and ydata. Enter the following example of a loglog plot.

```
> data := DataExample(power,exact): tdata := op(1,data): ydata := op(2,data):
> DoubleListPlot(map(log, tdata), map(log, ydata), title = `LogLog`);
>
```

What is the most striking feature of the loglog plot that you have just generated? Enter the power relation example commands a few more times--do the subsequent loglog plots continue to exhibit the "striking feature"? Place your answer below.
Answer:

What does a loglog plot look like for data which is only approximated by a power relationship? Let's examine some examples. Enter the following commands (a few times!) for examples.

```
> data := DataExample(power,approx): tdata := op(1,data): ydata := op(2,data):
> DoubleListPlot(map(log, tdata), map(log, ydata));
>
```

How would you describe the loglog plots that you just generated? Said another way, if a data set is well-approximated by a power curve, then what do you expect to find in its loglog plot? Place your answer below.

Answer:

¤ 3. Analyzing the Nature of a Data Set.

We can form semilog and loglog plots for any data set (discounting, if necessary, data points that would lead to logarithms of negative numbers). Let's see what these look like for data sets which are not well-approximated by exponential nor power relationships. Enter the following commands (a few times!) for examples.

```
> data := DataExample(norelation,approx): tdata := op(1,data): ydata := op(2,data):
> DoubleListPlot(tdata, map(log, ydata), title = `Semilog Plot`);
> DoubleListPlot(map(log, tdata), map(log, ydata), title = `Loglog Plot`);
>
```

Comment on the nature of the semilog and the loglog plots which you have generated. (Sometimes the "badness" of a fit can be missed, especially when a line seems to be "close" to all the data points. "Badness" might then be a "trend" in the data points which is not matched by the line, e.g., the data points seems to be leveling off while the line predicts continued growth. Look for such contradictory trends when analyzing the examples.) Place your comments below.
Answer:

From what you have observed thus far, how can semilog and loglog plots help you determine the nature of a data set? Place your answer below.
Answer:

¤ 4. Dr. Smith's Puppy.

Dr. David Smith, one of the two designers of Project CALC at Duke University, has a golden retriever named Sassafras. When Sassy was a puppy, Dr. Smith recorded her weight at ten day intervals:

birth (10/24/89)	3.25 pounds
10 days	4.25 pounds
20 days	5.5 pounds
30 days	7 pounds
40 days	9 pounds
50 days	11.5 pounds
60 days	15 pounds
70 days	19 pounds

We want to find a formula to model Sassy's weight as a function of age over the first six months of life. Then we can use this model to give (approximate) answers to questions like "How much did Sassy weigh at 55 days?" or "How much will she weigh at six months?" (The accuracy of the answers so obtained will be examined later.)

We must first enter the data into Maple. We will do this by creating two lists, days and pounds. (These will correspond to the lists tdata and ydata as used in previous sections.) We create the lists by simply writing out the values (enclosed in square brackets). Complete and Enter your definitions for days and pounds below.

```
> days := [ 1, 10, ??? ];
> pounds := [ ??? ];
>
```

We now plot our puppy growth data. This is most easily done by using the special command DoubleListPlot with the lists days and pounds. Complete and Enter your plotting command below.

```
> puppyPlot := DoubleListPlot( ??? , ??? ):";
>
```

• Maple Tip. You can improve the visual clarity of the current graph by adding "options" into the DoubleListPlot command. The option statements are placed following the two data lists in the DoubleListPlot command, and are all separated by commas, i.e., DoubleListPlot(tdata, ydata, option1=option1value, option2=option2value). Add options until the plot looks good to you. Here are some to try. (Remember: you can also change the size of the plot by using the mouse.) x=0 .. 20, title = `Your Plot's Title`. These options work in the same way as with the built-in Maple command plot.

¤ 5. Analyzing the Nature of Puppy Growth.

For this section you will need the lists days and pounds as defined in the previous section.

To further analyze the nature of Sassy's growth we will create semilog and loglog plots of the growth data. If one of these two graphs yields a straight line, then we can create a model function for Sassy's growth.

We first create a semilog plot for the growth data. The procedure is as in Section 1: take the original two lists, days and pounds, apply the Log function to the second of these lists (i.e., log(pounds)), then place them into a DoubleListPlot command. Carry out this procedure. (Add as many options into DoubleListPlot as are necessary to obtain a pleasing graph.) Complete and Enter the command below.

```
> semilogPlot := DoubleListPlot( ??? ):";
>
```

What conclusions do you draw from your semilog plot? Place your answer below.
Answer:

For the sake of completeness we also create a loglog plot for the puppy growth data. The procedure is similar to that used to create the semilog plot: take the original two lists, days and pounds, apply the log function to the both of them, then place new lists into a DoubleListPlot command. Carry out this procedure. (Add as many options into DoubleListPlot as are necessary to obtain a pleasing graph.) Complete and Enter the command below.)

```
> loglogPlot := DoubleListPlot( ??? );
>
```

What conclusions do you draw from your loglog plot? Place your answer below.
Answer:

Choose the plot which most looks like a line: semilogPlot or loglogPlot. You will fit a line to this plot by choosing values for the slope and the y-intercept of the line that best approximates the graph. Commands are given below to facilitate this process. Obtain initial estimates for slope and yIntercept from your plot above. Then complete and Enter the commands below. On the basis of the plot so produced, refine your estimates of slope and yIntercept and Enter the commands again. Continue in this way until you get a good linear fit. Complete and Enter the commands below--multiple times if necessary.

```
> Plot := ???:      # Choose semilogPlot or loglogPlot
> slope := ???;       # The slope of a good fitting line
> yIntercept := ???;     # The y-intercept of a good fitting line
> linePlot :=   plot(slope*t + yIntercept, t = 0..70):
> plots[display]({Plot, linePlot});
>
```

Assuming that you have gotten a good linear fit to your data, convert your line data (i.e., slope and yIntercept) into a function called linePuppy(t) by using the usual m*t + b form of a line. We'll use this in the next section to get a model function for Sassy's growth. Complete and Enter the following definition of linePuppy.

```
> linePuppy := t -> ???;
>
```

Note: The independent variable in linePuppy is t, not x.

¤ 6. Determining a Model for Puppy Growth.

For this section you will need the linear function linePuppy as defined in the previous section and the graphics puppyPlot as defined in Section 4.

• Maple Tip. You will need the exponential function. In Maple this is represented as either exp(t) or E^t (Maple uses E for the number e=2.7182...).

You ended the previous section by constructing a function, linePuppy(), whose graph is a line which approximates the data in the semilog plot. (We trust that you did choose the semilog plot in the previous section....) This means that the independent and dependent variables in the semilog plot are related (approximately) by the equation

dependent = linePuppy(independent).

However, in the semilog plot,

independent = t , and

dependent = log(weight)

where weight is Sassy's weight at t days after birth. Thus, substituting yields

log(weight) = linePuppy(t).

We ultimately desire to have a function ("model") for weight (puppy weight) in terms of t (time in days from birth). Well, it's a small step to go from the last equation to the desired function for weight. Carry out this last step. (Hint: How do you undo a "natural" logarithm?) Enter your definition for the model function below.

```
> weight := t -> ???;
>
```

Check your model weight(t) by plotting it along with the growth original data. The original data has already been placed into a plot, puppyPlot, which we can call on now. If the fit produced below is not good, then go back and check your work. Enter the following plotting command.

```
> weightPlot := plot(weight(t), t = 0 .. 70):
> plots[display]({puppyPlot, weightPlot});
>
```

Use your model to estimate Sassy's weight at 55 days. Is this a reasonable answer? If not, then how do you explain your model's "failure"? Place your computations and explanations below.)
Answer:

```
> evalf( weight( ??? ) );
>
```

Use your model to estimate Sassy's weight at three months. Is this a reasonable answer? If not, then how do you explain your model's "failure"? Place your computations and explanations below.)
Answer:

Use the model to predict Sassy's weight at six months. Is this a reasonable answer? If not, then how do you explain your model's "failure"? Place your computations and explanations below.)
Answer:

Comment on the following claim: "If a modeling function well-approximates a data set over a time interval 0 < t < t0, then it is likely that the model will produce accurate predictions for times greater than t0." Place your answer below.
Answer:

¤ EXERCISES.

• Exercise 1.

Suppose linePuppy was a linear fit to the loglog plot instead of to the semilog plot. Then how would you produce the modeling function weight from linePuppy, i.e., how would the equation for weight in Section 6 have to be changed?

```
> weight := t -> ??? ;
>
```

• Exercise 2.

The command DataExample has another option which produces "mystery" data sets, i.e., data sets which may exhibit an exponential relationship, a power relationship, or no simple relationship at all. Produce a few such data sets, and for each (1) determine if an exponential or a power relationship is present in the data, and (2) if an exponential or a power relationship is present, determine the parameters of a good fitting line for the semilog or loglog plot and construct a modeling function for the original data set. What follows is a summary of the commands to use for this exercise. Generate your data sets with the command:

```
> data := DataExample(mystery,`?`): tdata:=op(1,data): ydata:=op(2,data):
>
```

Then use DoubleListPlot to generate a (regular) plot of the data set, a semilog plot, and a loglog plot, naming these plots as shown below.

```
> dataPlot := DoubleListPlot( ??? , title=`Data`):";
> semilogPlot := DoubleListPlot( ??? , title=`SemiLog`):";
> loglogPlot := DoubleListPlot( ??? , title=`LogLog`):";
>
```

If the data set seems to exhibit an exponential or a power relation, then use the following command set from Section 5 to determine the best fitting line to either the semilog or the loglog plot.

```
> Plot := ???:             # Choose semilogPlot or loglogPlot
> slope := ???;            # Slope of a good fitting line
> yIntercept := ???;       # y-intercept of a good fitting line
> linePlot :=  plot(slope*t + yIntercept,t=start..finish):  # start=???, finish=???
> plots[display]({Plot,linePlot});
>
```

From the line data just obtained, define lineModel to be the linear function that best approximates the data in the semilog or loglog plot.

```
> lineModel := x -> ??? ;
>
```

Use lineModel to define model, a function that well-approximates the original data set. (You might need to consider the answer to Exercise 1.)

```
> model := x-> ??? ;
>
```

Test your model against the original data by modifying the plotting commands from Section 6:

```
> modelPlot := plot(model(t), t = start .. finish ):
> plots[display]({dataPlot, modelPlot});
```

LAB 5B: Rates of Change--Planetary Motion

□ NAMES:

□ PURPOSE.
The primary purpose of the laboratory will be to learn how to detect certain types of growth patterns in a given data set. The primary tools will be semilog and loglog plots. These tools will be applied to data on the distance from the sun and the length of a year for planetary objects in our solar system. The results will be used to construct a formula to model the relationship between these two quantities.

□ PREPARATION.
To fully understand the theory employed in this lab you should study Section 2.6, "Logarithms and Representation of Data," in Calculus: Modeling and Application. You should also review the Maple procedures that were discussed in Lab 1.

□ LABORATORY REPORT INSTRUCTIONS.
Note: Follow your instructor's instructions if they differ from those given here.

For this project you are not required to submit a formal report. Rather you are asked to submit a copy of this Worksheet with the questions answered in the indicated locations. You will be graded on this single submission. Write in complete sentences and be sure the sentences connect together in coherent ways.

□ MAPLE Procedures. Please Enter the following command.

```
> with(lab5);
> # If an Error is an output of the above line, see your lab instructor.
```

There are two procedures written for this lab. They are DoubleListPlot and DataExample. For more information on these use the ?DoubleListPlot and ?DataExample commands at the Maple command line.

□ PRELIMINARIES.

¤ 1. Data sets exhibiting exponential relationships.

A data set [[t1,y1], [t2,y2], ..., [tn,yn]] is said to exhibit an exponential relationship if there is an exponential function y = A*e^(k*t) such that yi = A*e^(k*ti) for every data point [ti,yi]. This is equivalent to observing that all the data points [ti,yi] lie on the exponential curve y = A*e^(k*t).

Enter the following command (a few times!) for examples.

```
> DataExample(exp,exact):
```

Rarely will real data, i.e., data measured from a real world phenomena, actually fall precisely on an exponential curve. The best we can hope for is that the data will be well-approximated by an exponential curve y = A*e^(k*t). The formula y = A*e^(k*t) is then said to be an exponential model for the relationship between the variables t and y.

Enter the following command (a few times!) for examples.

```
> DataExample(exp,approx):
>
```

¤ 2. Data sets exhibiting power relationships.

A data set [[t1,y1], [t2,y2], ..., [tn,yn]] is said to exhibit a power relationship if there is a power function y = A*t^r such that yi = A*(ti)^r for every data point [ti,yi]. This is equivalent to observing that all the data points [ti,yi] lie on the power curve y = At^r.

Enter the following command (a few times!) for examples.

```
> DataExample(power,exact):
>
```

Rarely will real data, i.e., data measured from a real world phenomena, actually fall precisely on a power curve. The best we can hope for is that the data will be well-approximated by a power curve y = A*t^r. The formula y = A*t^r is then said to be a power model for the relationship between the variables t and y.

Enter the following command (a few times!) for examples.

```
> DataExample(power,approx):
>
```

¤ 3. More general data sets.

We will first consider some data sets which are not well-approximated by either an exponential nor a power relationship. In such cases we can still try to fit exponential and power curves to the data, but the results are generally (but not always) bad. (Sometimes the "badness" of a fit will be overlooked at first, especially when a model curve seems to be "close" to all the data points. "Badness" might then be an obvious "trend" in the data points which is not matched by the model curve, e.g., the data points seems to be leveling off while the model curve predicts continued growth. Look for such contradictory trends when analyzing the examples.)

Enter the following commands (a few times!) for examples.

```
> DataExample(norelation,approx):
>
```

• Optional. Some of the examples you just generated will be pretty extreme: the attempted exponential and power function fits will be obviously terrible. However, some of the examples will be more subtle. Generate one of these "more subtle" examples and describe why you think each of the two proposed model curves are either good or bad. (Reread the parenthetical comment on "badness" in the previous paragraph.) Place your answers below.

Answer:

A reminder. Have you Saved your file yet? If not, do so now!

□ PROJECT.
¤ 1. Semilog Plots.

The output of the command DataExample consists of two lists tdata and ydata (the t coordinates and the y coordinates of the data points). These can be recombined in various ways by another new command: DoubleListPlot. Enter the following example.

```
> data := DataExample(exp,exact): tdata:=op(1,data): ydata:=op(2,data):
> DoubleListPlot(tdata,ydata);
>
```

The second plot is the output of the DoubleListPlot command. It is merely a scatter plot of the data set obtained by combining the t values of tdata with the corresponding y values from ydata (which is, in this simple case, the scatter plot of the original data set).

The primary value of DoubleListPlot is that it makes easy work of constructing semilog and loglog plots for a data set. We consider semilog plots now, and loglog plots later.

Recall that a semilog plot for a [t,y]-data set is merely a plot of the points [t,log(y)], i.e., take the logarithm of the y coordinate. Thus, to use DoubleListPlot to construct a semilog plot, we have only to take the logarithm of all the values in ydata. Enter the following example of a semilog plot.

• Maple Tip. In Maple the "ordinary" logarithm function, i.e., with base 10, is denoted by log[10](x), while the "natural" logarithm, i.e., with base e, is denoted by log(x). We will use the "natural" logarithm in what follows, although the process we now describe would work equally well with any choice of the base for the log function.

```
> data := DataExample(exp,exact): tdata:=op(1,data): ydata:=op(2,data):
> DoubleListPlot(tdata,map(log,ydata));
>
```

What is the most striking feature of the semilog plot that you have just generated? Enter the exponential example commands a few more times--do the subsequent semilog plots continue to exhibit the "striking feature"?

Place your answer below.
Answer:

What does a semilog plot look like for data which is only approximated by an exponential relationship? Let's examine some examples. Enter the following commands (a few times!) for examples.

```
> data := DataExample(exp,approx): tdata:=op(1,data): ydata:=op(2,data):
> DoubleListPlot(tdata,map(log,ydata));
>
```

How would you describe the semilog plots that you just generated? Said another way, if a data set is well-approximated by an exponential curve, then what do you expect to find in its semilog plot?

Place your answer below.

Answer:

¤ 2. Loglog Plots.

Recall that a loglog plot for a [t,y]-data set is merely a plot of the points [log(t),log(y)], i.e., take the logarithm of both the t and y coordinates. Thus, to use DoubleListPlot to construct a loglog plot, we have only to take the logarithm of all the values in both tdata and ydata. Enter the following example of a loglog plot.

```
> data := DataExample(power,exact): tdata:=op(1,data): ydata:=op(2,data):
> DoubleListPlot(map(log,tdata),map(log,ydata));
>
```

What is the most striking feature of the loglog plot that you have just generated? Enter the power relation example commands a few more times--do the subsequent loglog plots continue to exhibit the "striking feature"?

Place your answer below.

Answer:

What does a loglog plot look like for data which is only approximated by a power relationship? Let's examine some examples.

Enter the following commands (a few times!) for examples.

```
> data := DataExample(power,approx): tdata:=op(1,data): ydata:=op(2,data):
> DoubleListPlot(map(log,tdata),map(log,ydata));
>
```

How would you describe the loglog plots that you just generated? Said another way, if a data set is well-approximated by a power curve, then what do you expect to find in its loglog plot?

Place your answer below.

Answer:

¤ 3. Analyzing the Nature of a Data Set.

We can form semilog and loglog plots for any data set (discounting, if necessary, data points that would lead to logarithms of negative numbers). Let's see what these look like for data sets which are not well-approximated by exponential nor power relationships.

Enter the following commands (a few times!) for examples.

```
> data := DataExample(norelation, approx): tdata:=op(1,data): ydata:=op(2,data):
> DoubleListPlot(tdata, map(log, ydata), title=`Semilog Plot`);
> DoubleListPlot(map(log, tdata), map(log,ydata), title=`Loglog Plot`);
>
```

Comment on the nature of the semilog and the loglog plots which you have generated. (Sometimes the "badness" of a fit can be missed, especially when a line seems to be "close" to all the data points. "Badness" might then be a "trend" in the data points which is not matched by the line, e.g., the data points seems to be leveling off while the line predicts continued growth. Look for such contradictory trends when analyzing the examples.)

Place your comments below.

Answer:

From what you have observed thus far, how can semilog and loglog plots help you determine the nature of a data set?

Place your answer below.

Answer:

¤ 4. Planetary Years and Distance from the Sun.

The length of the year for planets in our solar system increases the farther the planet is from the Sun. Below is a list of lengths of years and corresponding mean distances from the Sun for the first six planets.

Planet	Distance from Sun (millions of miles)	Length of year (in Earth days)
Mercury	36	88
Venus	67	225
Earth	93	365
Mars	142	687
Jupiter	484	4333
Saturn	885	10759

We want to find a formula that gives the approximate length of a planet's year in Earth days as a function of its mean distance from the Sun. Then we can use this model to predict the length of a planet's year (or that of any object orbiting around the Sun) given is mean distance from the Sun.

We must first enter the data into Maple. We will do this by creating two lists, miles and days. (These will correspond to the lists tdata and ydata as used in previous sections.) We create the lists by simply writing out the values (enclosed in curly brackets). Complete and Enter your definitions for miles and days below.

```
> miles := [ 36, 67, ??? ];
> days := [???];
>
```

We now plot our planetary data. This is most easily done by using the special command DoubleListPlot with the lists miles and days. Complete and Enter your plotting command below.

```
> planetPlot := DoubleListPlot( ??? , ??? ):";
>
```

• Maple Tip. You can improve the visual clarity of the current graph by adding "options" into the DoubleListPlot command. The option statements are placed following the two data lists in the DoubleListPlot command, and are all separated by commas, i.e., DoubleListPlot(tdata, ydata, option1= option1value, option2=option2value). Add options until the plot looks good to you. (Remember: you can also change the size of the plot by using the mouse.) i.e. title=`Your Plot's Title` These options work in the same way as with the built-in Maple command plot.

¤ 5. Analyzing the Nature of the Planetary Data.

For this section you will need the lists miles and days as defined in the previous section.

To further analyze the nature of relationship between mean distance from the sun and the length of a planetary year we will create semilog and loglog plots of the data. If one of these two graphs yields a straight line, then we can create a model function for the relationship.

We first create a semilog plot for the data. The procedure is as in Section 1: take the original two lists, miles and days, apply the log function to the second of these lists (i.e., log(days)), then place them into a DoubleListPlot command. Carry out this procedure. (Add as many options into DoubleListPlot as are necessary to obtain a pleasing graph.) Complete and Enter the command below.

```
> semilogPlot := DoubleListPlot( ??? ):";
>
```

What conclusions do you draw from your semilog plot? Place your answer below.
Answer:

We now create a loglog plot for the planetary data. The procedure is similar to that used to create the semilog plot: take the original two lists, miles and days, apply the log function to both of them, then place the new lists into a DoubleListPlot command. Carry out this procedure. (Add as many options into DoubleListPlot as are necessary to obtain a pleasing graph.) Complete and Enter the command below.

```
> loglogPlot := DoubleListPlot( ??? ):";
>
```

What conclusions do you draw from your loglog plot? Place your answer below.
Answer:

Choose the plot which most looks like a line: semilogPlot or loglogPlot. You will fit a line to this plot by choosing values for the slope and the y-intercept of the line that best approximates the graph. Commands are given below to facilitate this process. Obtain initial estimates for slope and yIntercept from your plot above. Then complete and Enter the commands below. On the basis of the plot so produced, refine your estimates of slope and yIntercept and Enter the commands again. Continue in this way until you get a good linear fit. Complete and Enter the commands below--multiple times if necessary.

```
> Plot := ???:            # Choose semilogPlot or loglogPlot
> slope := ???:            # The slope of a good fitting line
> yIntercept := ???;     # The y-intercept of a good fitting line
> linePlot =   plot(slope*t + yIntercept, t=start..finish); #start=???, finish=???
> plots[display]({linePlot, loglogPlot});
```

Assuming that you have gotten a good linear fit to your data, convert your line data (i.e., slope and yIntercept) into a function called linePlanet(x) by using the usual m x + b form of a line. We'll use this in the next section to get a model function for the relationship in the planetary data. Complete and Enter the following definition of linePlanet.

```
> linePlanet := x -> ???;
>
```

¤ 6. Determining a Model for the Planetary Data.

For this section you will need the linear function linePlanet as defined in the previous section and the graphics planetPlot as defined in Section 4.

• Maple Tip. You will need the exponential function. In Maple this is represented as either exp(t) or E^t (Maple uses E for the number e=2.7182...).

You ended the previous section by constructing a function, linePlanet(...), whose graph is a line which approximates the data in the loglog plot. (We trust that you did choose the loglog plot in the previous section....) This means that the independent and dependent variables in the loglog plot are related (approximately) by the equation

dependent = linePlanet(independent).

However, in our specific loglog plot,

independent = log(x)
dependent =log(yearLength)

where yearLength is the length of a year (in Earth days) for a planet whose mean distance from the Sun is x million miles. Thus, substituting yields log(yearLength) = linePlanet(log(x)).

We ultimately desire to have a function ("model") for yearLength in terms of x (mean distance from the Sun). Well, it's a small step to go from the last equation to the desired function for yearLength. Carry out this last step. (Hint: How do you undo a "natural" logarithm?) Enter your definition for the model function below.)

```
> yearLength := x -> ???;
>
```

Check your model yearLength(x) by plotting it along with the original data. The original data has already been placed into a plot, planetPlot, which we can call on now. If the fit produced below is not good, then go back and check your work. Enter the following plotting command.

```
> yearLengthPlot := plot(yearLength(x), x = 0 .. 900):
> plots[display]({planetPlot, yearLengthPlot});
>
```

¤ 7. Predicting Other Planetary Years.

For this section you will need the modeling function yearLength as defined in Section 6.

The mean distances of the three planets farthest from the sun are

Uranus:	1782 million miles,
Neptune:	2790 million miles,
Pluto:	3660 million miles.

Use the formula you obtained in Section 6 to estimate the length of the year on each of these three planets. The observed length of the year on each is 30685 days, 60188 days, and 90700 days respectively. Compare these numbers to your predicted values. Enter your computations below.)

Answer:

```
> evalf( ???( ??? ) );
> ???
>
```

The asteroid Ceres has a mean distance from the sun of 257 million miles. Estimate the length of the year on Ceres, and compare to the observed value of 1681 days. Enter your computations below.)

Answer:

¤ EXERCISES.

• Exercise 1.

Suppose linePlanet was a linear fit to the semilog plot instead of to the loglog plot. Then how would you produce the modeling function yearLength from linePlanet, i.e., how would the equation for yearLength in Section 6 have to be changed?

Answer:

```
> yearLength := x-> ??? ;
>
```

• Exercise 2.

The command DataExample has another option which produces "mystery" data sets, i.e., data sets which may exhibit an exponential relationship, a power relationship, or no simple relationship at all. Produce a few such data sets, and for each (1) determine if an exponential or a power relationship is present in the data, and (2) if an exponential or a power relationship is present, determine the parameters of a good fitting line for the semilog or loglog plot and construct a modeling function for the original data set. What follows is a summary of the commands to use for this exercise.

Generate your data sets with the command:

```
> data := DataExample(mystery,`?`): tdata:=op(1,data): ydata:=op(2,data):
>
```

Then use DoubleListPlot to generate a (regular) plot of the data set, a semilog plot, and a loglog plot, naming these plots as shown below.

```
> dataPlot := DoubleListPlot( ??? , title=`Data`):";
> semilogPlot := DoubleListPlot( ??? , title=`SemiLog`):";
> loglogPlot := DoubleListPlot( ??? , title=`LogLog`):";
>
```

If the data set seems to exhibit an exponential or a power relation, then use the following command set from Section 5 to determine the best fitting line to either the semilog or the loglog plot.

```
> Plot := ???:          # Choose semilogPlot or loglogPlot
> slope := ???:          # Slope of a good fitting line
> yIntercept := ???:       # y-intercept of a good fitting line
> linePlot :=  plot(slope*t + yIntercept,t=start..finish):
> plots[display]({Plot,linePlot});
>
```

From the line data just obtained, define lineModel to be the linear function that best approximates the data in the semilog or loglog plot.

```
> lineModel := x -> ??? ;
>
```

Use lineModel to define model, a function that well-approximates the original data set. (You might need to consider the answer to Exercise 1.)

```
> model := x-> ??? ;
>
```

Test your model against the original data by modifying the plotting commands from Section 6:

```
> modelPlot := plot(model(t), t = start .. finish ):
> plots[display]({dataPlot, modelPlot});
>
```

• Exercise 3.

Dr. David Smith, one of the two designers of Project CALC at Duke University, has a golden retriever named Sassafras. When Sassy was a puppy, Dr. Smith recorded her weight at ten day intervals:

birth (10/24/89)	3.25 pounds
10 days	4.25 pounds
20 days	5.5 pounds
30 days	7 pounds
40 days	9 pounds
50 days	11.5 pounds
60 days	15 pounds
70 days	19 pounds

We want to find a formula to model Sassy's weight as a function of age over the first six months of life. Then we can use this model to attempt answers to questions like "How much did Sassy weigh at 55 days?" or "How much will she weigh at six months?" What follows is a summary of the commands to use for this purpose.

Define your data sets:

```
> days := [ ??? ];
> pounds := [ ??? ];
>
```

Use DoubleListPlot to generate a (regular) plot of the data set, a semilog plot, and a loglog plot, naming these plots as shown below.

```
> puppyPlot := DoubleListPlot( ??? , title = `Data`):";

> semilogPlot := DoubleListPlot( ??? , title = `SemiLog`):";

> loglogPlot := DoubleListPlot( ??? , title = `LogLog`):";
>
```

If the data set seems to exhibit an exponential or a power relation, then use the following command set from Section 5 to determine the best fitting line to either the semilog or the loglog plot.

```
> Plot := ???:    # Choose semilogPlot or loglogPlot
> slope := ???;      # Slope of a good fitting line
> yIntercept := ???;   # y-intercept of a good fitting line
> linePlot :=   plot(slope*t + yIntercept,x= ??? .. ???):
> plots[display]({ Plot, linePlot });
>
```

From the line data just obtained, define linePuppy to be the linear function that best approximates the data in the semilog or loglog plot.

```
> linePuppy := t -> ??? ;
>
```

Use linePuppy to define weight(t), a function that well-approximates Sassy's weight at t days from birth. (You might need to consider the answer to Exercise 1.)

```
> weight := t -> ??? ;
>
```

Test your model for Sassy's weight against the original data by modifying the plotting commands from Section 6:

```
> weightPlot := plot(weight(t),  t = ??? .. ???):
> plots[display]({puppyPlot, weightPlot});
>
```

Use your model to estimate Sassy's weight at 55 days. Is this a reasonable answer? If not, then how do you explain your model's "failure"? Place your computations and explanations below.)

Answer:

```
> evalf( ??? );
>
```

Use your model to estimate Sassy's weight at three months. Is this a reasonable answer? If not, then how do you explain your model's "failure"? Place your computations and explanations below.)

Answer:

Use the model to predict Sassy's weight at six months. Is this a reasonable answer? If not, then how do you explain your model's "failure"? Place your computations and explanations below.)

Answer:

Comment on the following claim: "If a modeling function well-approximates a data set over a time interval $0 < t < t0$, then it is likely that the model will produce accurate predictions for times greater than t0." Place your answer below.)

Answer:

LAB 6: Raindrops

□ NAMES:

□ PURPOSE.
In this lab we will study Initial Value Problems: problems which require us to find a particular solution to a differential equation, the particular solution which satisfies a given initial condition. The initial condition is usually the starting state of a system (e.g., "The velocity of the ball before we dropped it was zero."), and the desired solution is the state of the system at any subsequent time t (e.g., "The velocity of the ball was 32.2t feet per second at t seconds after the ball was dropped."). The problems considered in this lab model falling bodies-raindrops in particular-subject to air resistance.

We will use Euler's Method, a numerical method for approximating the solutions of initial value problems. Although not identified by name, this method has been used in previous labs.

□ PREPARATION.
Review the discussions in Calculus: Modeling and Application of the falling body problem (Sec. 3.3 in Chapter 3) and the Chapter 3 Lab Reading: Raindrops. In addition, read these instructions carefully before coming to the lab, and solve the exercise given at the end of Section 1 of the Preliminaries.

□ Maple Procedures.
Be sure to initialize the worksheet before beginning the lab by activating the following command.

```
> with(lab6);
> # If an Error is an output of the above line, see your lab instructor.
```

□ LABORATORY REPORT INSTRUCTIONS.
Note: Follow your instructor's instructions if they differ from those given here.

For this project you are not required to submit a formal report. Rather you are asked to submit a copy of this Worksheet with the questions answered in the indicated locations. Only one submission will be made. Write in complete sentences and be sure the sentences connect together in coherent ways.

□ PRELIMINARIES.
In this section you will be guided through a process of discovering the formulas that are needed in this lab.

• Learning Tip.
Writing important things in your own handwriting is a good way to learn them, often better than typing. Consequently, we ask you to carefully write down the basic assumptions of this next section, i.e. F=m*a, a=dv/dt, ... , as well as the answers to the following set of questions. They will lead you in a gentle way to discovering the important formulas that will allow us to solve some practical problems.

¤ 1. A model for falling bodies--no air resistance.

All our models for the velocity of a falling body will be based on Newton's Second Law of Motion, which states that force equals mass times acceleration, i.e., F = m*a. Here F is the force exerted on an object of mass m, causing the object to have an acceleration a. However, acceleration a is defined to be the derivative of the velocity v, i.e., a = dv/dt. Thus Newton's Law can be rewritten as

F = m*dv/dt

The primary force on a falling body is gravity, the pull of the Earth's mass on the object. Our first model for a falling body will consider gravity to be the only force on the object. It is known through experimental observation that (near the surface of the Earth) the force of gravity on an object is proportional to the mass of the object, i.e., there is a constant g such that

F = m*g

(By experimentation the value of the constant g is known to be g = 32.2 ft/sec^2.) Equating our two formulas for the force F and dividing by m yields a differential equation. Write down the differential equation on paper and the steps leading to it, then check your answer by activating the next Maple command.

```
> Ans(1);
> # Scroll down with the down arrow.
```

If we assume that our object was initially at rest at time t=0, then we obtain an initial condition of v(0)=0, i.e., at time t=0 the velocity is v=0. Combining this condition with the differential equation yields our first Initial Value Problem (for the velocity function v=v(t)).

Write down the IVP on paper, then check your answer by activating the next Maple command.

```
> Ans(2);
```

It is easy to guess the solution to this problem:
Write down the solution on paper, then check your answer by activating the next Maple command.

```
> Ans(3);
```

However, this solution generates a second differential equation: since the velocity v is itself defined to be the derivative of the distance fallen s, i.e., v = ds/dt, then what do you get when you substitute this into our formula for v ? Write down the differential equation of paper, then check your answer by activating the next Maple command.

```
> Ans(4);
```

Since the distance fallen is s=0 when t=0, then we obtain an initial condition of s(0)=0. Combining this condition with the differential equation yields our second Initial Value Problem (for the distance fallen function s=s(t))... Write down the IVP on paper, then check your answer by activating the next Maple command.

```
> Ans(5);
```

As was the case with the first IVP, it is easy to guess the solution to this problem: Write the solution on paper then check it.

```
> Ans(6);
>
```

¤ Exercise. Using the model just devised, compute how long (in seconds) it would take a raindrop to fall from a height of 3000 feet. How fast would it be traveling when it hit the ground? Give your answer first in feet per second, and then convert it to miles per hour. Would you want to get hit with a raindrop traveling at this speed?? Place your answers below.

• Maple Tip. The fsolve command might be handy for this exercise:

```
> fsolve( ...=... , t);
>
```

Answers:

¤ 2. A model for falling bodies-with air resistance.

In this lab we will investigate a more sophisticated model for a falling body, one that takes into account the resisting force of the air through which the object falls. The usual physical assumption is that the force of air resistance is proportional to some power of the velocity, but the particular power (first, second, other) depends on the particular object.

We consider raindrops falling from a cloud 3000 feet above the ground. If the raindrop is small, say a drop of diameter 0.00025 feet (or 0.003 inches, a size found in a drizzle), the force of air resistance is modeled well by a multiple of the first power of the velocity; in other words, this force can be described by Fr = -k*v for some constant k. (The minus sign indicates that the force is in the opposite direction to the velocity, i.e., upward rather than the positive downward direction.) When combined with the force of gravity, Fg = m*g, this yields the total force F = Fg + Fr on the raindrop:

Write down the total force on paper, then check it.

```
> Ans(7);
>
```

We again recall Newton's Second Law of Motion:

F = m*dv/dt

Equating our two formulas for the force F and dividing by m yields what differential equation? Write down the differential equation on paper and on paper, then check it.

```
> Ans(8);
>
```

We'll let c represent the quotient of the two constants k and m. When we attach our initial condition, v(0)=0, we obtain our (third) Initial Value Problem. Write down the IVP on paper and on paper, then check it.

```
> Ans(9);
>
```

Experimental evidence gives an approximate value of 52.6 sec^(-1) for c.

We want to know how, under these assumptions, the velocity varies as a function of time. Our first impulse might be to copy the solution method of Section 1 (where we dealt with the simpler problem of no air resistance): guess a formula for v as a function of time, then use this formula to draw the graph (velocity verses time) and calculate the desired velocity values.

Well yes, this can be done-but "guessing" a formula for v as a function of time is more difficult for our new problem than it was for our old. (Systematic ways to solve this type of IVP are considered in both Chapters 3 and 7 of Calculus: Modeling and Application.) We will, however, take a different approach in this lab and use a numerical technique called Euler's Method. Although Euler's Method will only approximate the desired solution, it has the distinct advantages of applicability to any initial value problem (independent of whether we can "guess" a formula for an exact solution) and relative ease of understanding and use (of course, "ease of understanding and use" can have different meanings for the novice calculus student as opposed to the instructor).

¤ 3. Euler's Method-Geometric Description.

Recall the Initial Value Problem that we need to solve:

Find v=v(t) such that dv/dt = g - c*v when v(0)=0.

Also recall that, we have specific values for g and c (both obtained experimentally):

g = 32.2 and c = 52.6

We will calculate approximate values for the velocity v at n equally spaced points in some fixed time interval. The initial value for n will be 20, although we will increase this number later in the lab. The time interval we use is 0 <= t <= 0.2 -- the reason for this choice will be explained below. Thus the distance between consecutive t values will be Dt = Delta = 0.2/n . (We would prefer to use the Greek letter, capital delta, (a triangle) in place of the D in Dt. This is not convenient to do in the current version of Maple. The symbol we would prefer to use can be viewed by activating the next line.)

```
> Delta;
> # Scroll down with the down arrow.
```

Thus, our goal is to estimate the velocity at times

t(0)=0, t(1)=Dt, t(2)=2Dt, t(3)=3Dt, ... , t(n)=nDt.

Our estimated velocity values at these times will be denoted by

v(0), v(1), v(2), v(3), ... , v(n)

Our method for estimating the velocity values will be recursive, i.e., v(k) will be calculated from the previous v(k-1) for each k=1,2,3, ... To begin the calculation, how do we obtain v(1) from v(0), the initial velocity? We will answer this in a geometric fashion (which should remind you of the earlier laboratory on Limited Population Growth). (Note: Although in our case v(0) is zero, for ease of generalization, our pictures will show a non-zero initial velocity v(0).)

We look at the graph of velocity verses time on the (t,v) coordinate plane. The first diagram ...

...shows a graph of the starting situation: the initial velocity v(0) is shown as a vertical line segment of length v(0) at the starting time t(0)=0. We now add the...

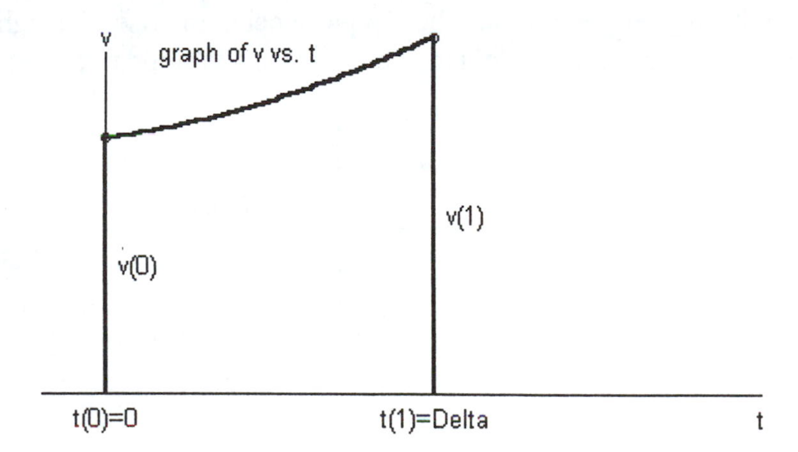

...graph of v verses t. Our next velocity value, v(1), is the length of a vertical line segment at t(1). However, the value of v(1) is not known to us, and hence we will estimate it. We do this by...

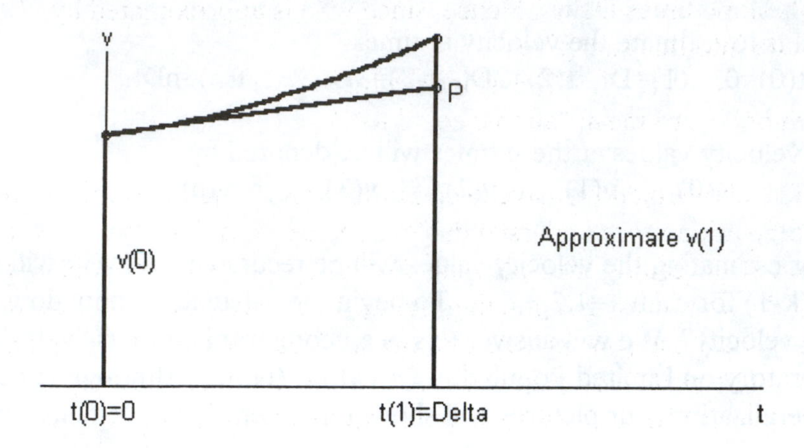

...drawing the tangent line to the graph at t=t(0). Follow this tangent line to the point P (as shown above)-P is the top of a vertical segment that approximates v(1). Moreover, we can compute the length of this new line segment by breaking it into two pieces...

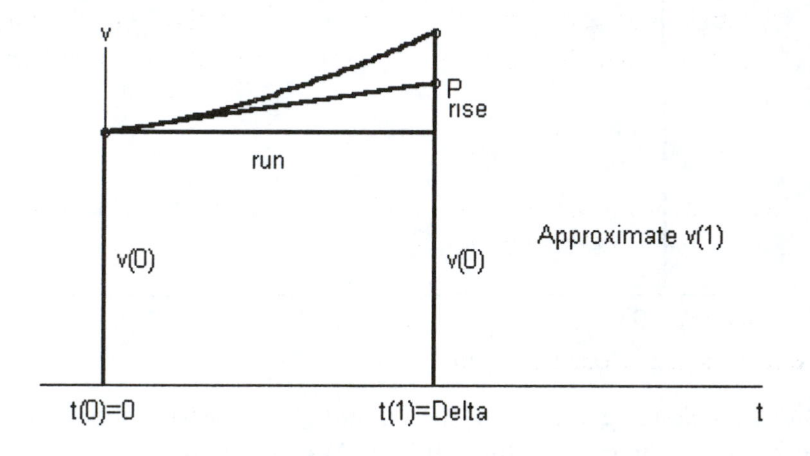

...the bottom piece having length v(0) and the top piece being the rise of a right triangle with run = Delta. Using slope = rise / run, the defining equation for the slope of a line, we see that...

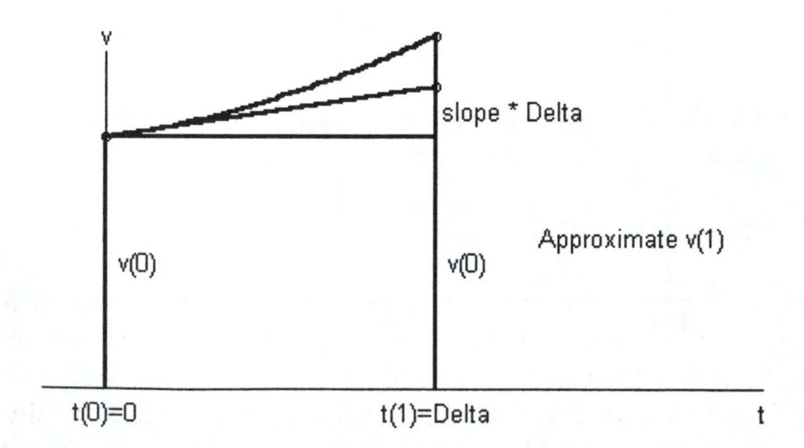

...the rise equals slope times Delta. Hence, since v(1) is approximated by v(0) + rise, substituting rise = slope*Delta into this sum gives

v(1) ~ v(0) + slope*Delta.

(We use the symbol ~ to mean "almost equal to".)

• Maple Tip. The above pictures have been put into an animation (in the hopes that it will make the preceding derivation more understandable). To see the animation, activate the following Maple command after reading the following discussion. The lower control panel starts with a square (stop), followed by a right pointing triangle (run animation), followed by an arrow pointing into a vertical line (one picture at a time). You can restart the animation by clicking on the 1:1 button. The next two commands either slow down or speed up the animation, and the last two on that line can be toggled back and forth; experiment with them.

```
> Euler(now );
>
```

The last equation (v(1) ~ v(0) + slope*Delta) is the key for Euler's Method of approximating the solution of an initial value problem. It's valuable because the slope (of the tangent line) equals the derivative dv/dt, which is given by our original differential equation when t=t(0) and v=v(0):

slope := dv/dt = g - c*v(0).

Substituting this value of the slope into the previous Euler's Method equation yields

v(1) ~ v(0) + (g - c*v(0))*Delta .

Great! This gives us a method for going from v(0) to v(1). But how do we go from v(1) to v(2)? Easy, we use the same equation, only with v(0), v(1) replaced by v(1), v(2):

v(2) ~ v(1) + (g - c*v(1))*Delta .

In general, to go from v(k-1) to v(k) we have ...
Write down the recursion equation on paper and below, then check it.

```
> Ans(10);
>
```

This equation, along with the initial value v(0)=0 and the assignment of values to the step size Delta, plays the central role in our computations.

¤ 4. Euler's Method-Algebraic Description.

Euler's Method is so central to the course that we will give another "derivation" for its key equation, this time algebraic rather than geometric. (Seeing a topic from more than one viewpoint can enhance understanding.) As is often the case, the algebraic derivation is quicker than the geometric-in this sense it is easier to follow since there is less to follow!-but the algebraic approach can often seem like symbol pushing, devoid of real meaning. The algebraic and geometric viewpoints in this sense complement each other.

We start with the differential equation, evaluated at t = t(k-1), v = v(k-1):

dv/dt = g - c*v(k-1),

along with the derivative approximation given by

dv/dt ~ Dv/Dt ~ (v(k) - v(k-1))/Dt .

Equating these two expressions for dv/dt yields...
Write down the equation on paper, then check it.

```
> Ans(11);
>
```

Solving for v(k) results in the equation...

Write down the equation on paper, then check it.

```
> Ans(12);
>
```

This agrees with what we obtained at the end of the geometric derivation in the previous section.

□ PROJECT.
¤ 1. Implementing Euler's Method.

We begin with a "manual" calculation (to be sure that you understand the Euler's Method procedure used throughout the lab). We assume a small rain drop begins its fall at time t(0)=0 with an initial velocity of v(0)=0. We know the two constants appearing in the differential equation governing the fall have values g=32.2 ft/sec^2 and c=52.6 sec^(-1). We will increment our times by Delta = Dt = 0.01 sec. Given these values and the formulas from Preliminaries, fill in the missing values for the following table (the values on the last row of the table are provided as checks on your work--if you don't obtain numbers very close to them, you need to rethink your computations):

Complete the table below--you can use the Maple input lines below the table.

Time	Velocity
t(0)= ??	v(0)= ??
t(1)= ??	v(1)= ??
t(2)= ??	v(2)= ??
t(3)= 0.03	v(3)= 0.546974

```
> Delta :=.01:   v(0) := 0:
> v(1) :=  v(0) + (32.2- 52.6*v(0))*Delta;   # Replace indices to obtain v(2) & v(3).
>
```

• Maple Tip.
The above method of writing v(k), for some particular k, in terms of v(k-1) only works when specific numbers have been inserted. It does not work as a general definition for v(k) in terms of v(k-1). The best way to do that is to use the Maple arrow or proc (procedure) command and will be done later.

We now automate the process that you just implemented manually.

Enter the values of the constants g=32.2 and c=52.6 and the initial velocity v(0)=0.

Type and Enter the commands on the line below.

```
>  g:= ?? :   c:= ?? :
>
```

We examine the time period from 0.0 to 0.2 seconds (reasons for this choice will become clear in a moment), and we will do so using a variety of time intervals Delta. To do so we will simply specify n, the number of (equal) time intervals that we wish to create from the time interval (0.0, 0.2). Then Delta=0.2/n. Input the (initial) value n=20 and the equation for Delta.
Type and Enter the commands on the line below.

```
>  n:= ?? : Delta:= ?? ;
>
```

Now define the time values, t(0), t(1), ..., t(n). As a preliminary task, fill in the appropriate values in the following list (remember that the distance between consecutive t values is delta):

t(0) := ??
t(1) := ??
t(2) := ??
t(3) := ??

Given the entries in your list, do you see what the general formula should be, i.e., the formula for t(k) in terms of k and Delta? Type it here:

t(k) = ??? for k = 0..n.

Complete and Enter the following Maple definition for this function.

```
> t := k -> ???
```

To check your definition, you can activate the following line.

```
> seq( t(k),k=0..n);
>
```

We are now ready for the major step: the command defining the velocity values v(1), v(2), ... , v(n) at the time values t(1), t(2), ... , t(n) (the value v(0) is not included here because it is given, not computed). The equation to use for v(k), as k goes from 1 to n, was derived in the Preliminaries section (in fact, derived twice!). Since it is recursively defined, it is best to use a Maple proc.

Complete the following definition.

```
> v := proc(k)          # This is in the form that gives the option remember.
>      v(k) := ???      # Same formula here as earlier but in a
>      end;             # proc command.
> v(0) := 0;            # Initial value condition; needed.
```

First test it by activating the following command.

```
> v(4);
>
```

You should get a decimal number starting with .58... . If you get something else, (for example a long expression with several c's and g's) , go back and fix the problem (activate the command initializing c and g).

Warning!!! Next, It is important to test a recursion relationship before running it too many times. For that reason you will first test your seq command with k running from 1 to 4.

Don't forget to save your work.

Type and Enter the "test" seq command on the line below.

```
> seq(v(k), k=1..4);
> # Scroll down with the down arrow.
```

Important!!! Your result should be a list of four numbers which match values you computed by hand at the start of this section. If anything else appears, then do not execute the seq command for k larger than 4 until you fix the problem!!

If you have a problem... Note the names of any non-numerical symbol which appears in your output (e.g., v(0), cv(0), c, etc.). One of the following problems has probably occurred:
(1) the symbol should have been assigned a value in the previous steps (which for some reason did not happen) or
(2) the symbol was incorrectly introduced by an error in the statements you typed. These two problems most often occur for the following reasons:

(1) You did not Enter the statement which assigned a value to your symbol. If so, then go back, find the necessary statement, and Enter it.

(2) You may have forgotten that multiplication is designated with a *.
For this reason c*v(k-1) denotes the multiplication of c and v(k-1). But cv(k-1) does not indicate a multiplication-it indicates a new indexed variable cv. Correct any errors of this nature in your seq commands. Continue to correct and reenter your seq (with k going up only to 4) until the resulting output is a list of four pairs of numbers. Correct and reEnter your "test" seq command above as needed.

When all problems have been corrected... You are ready to run your seq command for the full range of k values. Simply change k=1..4 to k=1..n, and then Enter the command.

Enter your seq command with the full k range.

```
> seq(v(k), k=0..n);
>
```

We now create a list of data points [t,v] for velocity v(k) as a function of time t(k) by using the seq command. We call this data list tvData(n) (remember: the current value of n is 20) It will be analyzed in the next section.

• Maple Tip.
Again, the following way of defining tvData(n) only works for a preassigned value of n. Later, to define tvData(n), for general n, we will use the Maple arrow command.

Enter the command shown on the line below.

```
> tvData(n) := [ seq([t(k),v(k)], k=0..n) ];
> # Scroll down with the down arrow.
```

The output of tvData(n) should be pairs of numbers. If any symbols appear, then go back and check your prior work.

¤ 2. Executing Euler's Method.

We create a scatter plot of the data file tvData(n) (velocity as a function of time) by using the following plot command. We call the output tvPlot(n).

Enter the plotting command on the line below.

```
> plot(tvData(20), style=POINT, symbol=CIRCLE, color=blue);
>
```

There is a big surprise in this graph, something that did not occur in the model without air resistance as discussed in Preliminaries. Describe what this surprise is.
Answer:

We will recompute the data set using larger n values (we expect larger values of n to yield more accurate results--why?).

One way to do this: find the statement n=20 in the previous section, change it to n=40, and then re-execute it and the statements which followed.

• Maple Tip. It is particularly important to re-execute the definition of the recursive function v. Why?

```
> Ans(13);
>
```

For another, more sophisticated, way to do it we have collected the relevant commands from the previous section and placed them below. Activate the regions.

```
> g := 32.2:  c := 52.6:  n := 'n':  v:='v':
> t := k -> Delta*k:               # Delta has not been defined yet; that's okay.
> readlib(forget):                 # Needed in the tvData definition.
>
> v := proc (k)                                   # Option remember is invoked.
>     v(k) := v(k-1) +(g - c*v(k-1))*Delta;
>     end:
>
> tvData := proc (n)  local k; global Delta; option remember;
>      Delta := .2/n;             # Delta is now assigned a value.
>      forget(v);                 # This line and the next re-initialize the remember
>      v(0) := 0;                 #  table for v so that it is valid for the new Delta.
>      [seq( [t(k),v(k)], k=0..n)];
>      end:
```

Before engaging in serious use, let's test tvData(n).
Enter the following commands.

```
> tvData(5);
>
```

Important!!! Your result should be a list of six pairs of numbers where the last visible number is 1.61... . If anything else appears, then do not execute the tvData(n) command for n larger than 5 until the problem is fixed!!

Your work in constructing the tvData(n) command will now be rewarded. Compute the Euler's Method data for n=20 by Entering the statement tvData(20).
Type and Enter your statement below.

```
> tvData(20);          # And Enter the next line.
> tvData(40);          # If you don't want to see the output, replace ; by : .
>
```

• Maple Tip.
Notice how all of the Euler's Method steps are now executed by activating just the one region containing the definition of tvData. The most intensive computation for this is in computing the recursively defined v(k), but this function uses a memory table, for a fixed n, so that it can efficiently compute the recursively defined v(k). Likewise, tvData has its own memory table so that once it computes a table of values, say tvData(40), it will store it and if, in the same session, it is called on again to recall that data, it just reads it out of the memory table, which is very efficient. For those of you who are interested, further information on this will be given at the end of this worksheet, in the Addendum entitled Maple Explanations.

We are now ready to plot the data for different values of n. Here we engage in more Maple magic. The idea of the first function below is to assign different color graphs to different values of n. RGB stands for red, green, and blue. For example color(RGB, 1.,0,0) is red.

Activate the following region.

```
> color := n -> COLOR(RGB,          # Different values of n to different colors.
>      max(1-.001*(n-60)^2,0), (n/200)^2, 1-n/200.0):
> tvPlot := n ->
>      plot(tvData(n), style=POINT, symbol=CIRCLE, color=color(n)):
```

Try it by activating the following command.

```
> tvPlot(20);
>
```

How did that work?! Fix it if necessary. Well, the command you Entered earlier to plot the n=20 data was very similar to this but worked only for a value of n that had already been assigned. Here we define tvPlot to be a function of n, for any n. In fact, Entering tvPlot(n) for any specific value of n causes tvData(n) to be computed, or looked up in the remember table, and plotted.

We will now show both of our data sets (for n=20 and n=40) together on the same plot. This is done with a plots[display] command.
Enter the following statement.

```
> plots[display]([tvPlot(20),tvPlot(40)]);
>
```

Briefly describe the differences that you see between the two data plots. Place your observations below.
Answer:

You now have the tools to quickly and efficiently execute Euler's Method for our small raindrop (the function tvData(n)), plot the results (tvPlot(n)), and compare the plots for a variety of n values. We now can apply these tools to analyze the raindrop problem. This is what we do in the subsequent sections.

¤ 3. Terminal Velocity.
Display plots for the small raindrop problem for 4 or 5 values of n between 20 and 200. (Suggest 20, 40, 80, 130, 200.) Then show all of the graphs in the same plot. Complete and Enter the plotting commands below.

```
> plots[display]([tvPlot(???),tvPlot(???), ???]);
>
```

Describe the sequence of plots and how they change as n increases. Describe how the velocity of the falling object varies as time increases. Place your answer below.

Answer:

Use your calculated velocity values to estimate to four decimal places the limiting value of the velocity as time increases. This is called the terminal velocity. Express your answer in both feet/sec and miles/hour. (Hint: For example, you can see all of tvData(100) by simply Entering the statement tvData(100). Alternately, to see just the last data point, you have only to Enter the statement last(tvData(100)). The "last" command was written specifically for this lab.

Place some calculations below.

```
> last(???);
>
```

Place your answer to the question: What is the terminal velocity of small rain drops, to four decimals, in both feet/sec and miles/hour?

Answer:

Compare your terminal velocity with what you obtained in the Preliminaries as the velocity for a small raindrop when it hits the ground after falling 3000 feet. (Prepare to be surprised!) Does it seem reasonable to disregard air resistance in models for falling objects?

Answer:

As t increases and the velocity v approaches the terminal velocity, what happens to the slope of the velocity verses time curve? Hence what happens to the derivative dv/dt ?
Answer:

Using your answer to the preceding question, calculate the terminal velocity directly from the original differential equation, $dv/dt = g - c\,v$.(Hint: What happens to each side of the differential equation as t becomes large?)

Place your answers below.
Answer:

As you have seen, the small raindrop approaches its terminal velocity quite rapidly. Estimate the time it takes the drop to fall to the ground from a starting height of 3000 feet by assuming that the velocity is the constant terminal velocity during the whole duration of the fall. How does this time compare to your answer to the Exercise in the Preliminaries (where no air resistance was assumed)? (This is a trivial computation, but prepare to obtain a surprising result.)
Computations and Answer:

¤ 4. A Closed Form Solution.

For this particular initial value problem (modeling a small raindrop), we can find a formula for the solution:

V(t) = (g/c) (1-e^(-ct))

In Chapters 3 and 7 of Calculus: we will discuss how you could find this solution. For now, Enter this into the worksheet as a Maple function V(t).

Enter your definition for V(t) below.

• Maple Warning. Enter V(t) in proper Maple notation. The definition given above for V(t) is not proper Maple notation. Common mistakes come from breaking the following rules:
(1) ct is not the same as c*t;
(2) Maple knows E; it does not know e;
(3) algebraic grouping can only be done with (...), never with [...].

Complete and activate the following line. Then check it.

```
>  V:= t -> ???;
> Ans(14);
```

Plot this solution function for t values between 0 and 0.2 and call the result Vplot.

```
> Vplot := plot ???
>
```

• Maple Tip. If f is a function defined in the functional notation f := x -> f(x), then to plot f for x in [a,b], use the command plot(f, a..b); . If b:= .2 you can use a.. .2 .

Now, plots[display] this graph together with tvPlot(20). Repeat this with an Euler's Method plot for some higher value of n.

Enter your computations below.

```
>  plots[display]([???, ???]);
>
```

• Maple Tip. plots[display]([plot1, plot2]); displays both plot1 and plot2 on the same graph.

How well do the Euler's Method approximations compare with the exact solution?
Answer:

Our "closed form" solution gives us yet a third method for computing the terminal velocity v0. Use the formula for V(t) to compute v0. (Hint: What happens to e^(-ct) as t gets large?) Place your calculations below.
Answer:

¤ 5. Modeling Large Raindrops.

For large raindrops, say with diameter 0.004 feet (or 0.05 inches, a size typical of drops in a thunderstorm), the force of air resistance is better modeled as a multiple of the square of the velocity. The differential equation now has the form

dv/dt = g - a*v^2

where a is another constant. In this case, the experimental evidence yields a value for a of 0.115. This becomes an initial value problem upon the assignment of v(0)=0.

We wish to use Euler's Method to approximate the solution to this new IVP, this time over the time interval 0 <= t <= 2.0. The Maple statements needed to do this computation are easy variations of the statements used in the small raindrop model. However, to save time we give the statements below, with changes indicated by # comments.

Save your worksheet and then activate the following regions.

```
> g := 32.2:  a := 0.115:  n := 'n':  v:='v': # c:=52.6 has been changed to a:=0.115
> t := k -> Delta*k:
> v := proc (k)  v(k) := v(k-1) +Delta*(g - a*v(k-1)^2);end: # c*v(k-1) to  a*v(k-1)^2
> readlib(forget):
>
> tv2Data := proc (n)  local k; global Delta; option remember;
>                                                     # tvData to tv2Data
>     Delta := 2.0/n;                                 # .2/n  to  2.0/n
>     forget(v);
>     v(0) := 0;
>     [seq( [t(k),v(k)], k=0..n)];
>     end:
>
> color := n -> COLOR(RGB,
>     max(1-.001*(n-60)^2,0), (n/200)^2, 1-n/200.0):
>
> tv2Plot := n ->                                 # tvPlot  to  tv2Plot
>     plot(tv2Data(n), style=POINT, symbol=CIRCLE, color=color(n)):
```

You should check these with some small numbers. Activate the next line.

```
> tv2Data(5);
```

The output should be six pairs of numbers. If not, trouble-shoot your work.

```
> tv2Plot(20);
```

Graph the data sets for n=20 and n=100. Then plots[display] the two graphs together in one plot. Complete and Enter the graphing command below.

```
> plots[display]([tv2Plot(??), tv2Plot(??)]);
>
```

Describe the differences and similarities between the results for the large raindrops modeled here and the small raindrops modeled earlier. Place your answers below.
Answer:

Estimate the terminal velocity using the data computed in answer to the previous question. (Hint: For example, to see the last data point in tv2Data(100), you have only to Enter the statement last(tv2Data(100)).) Place your answer below.

```
> last(tv2Data(5));
>
```

Answer:

Estimate the terminal velocity by a second method, using the differential equation
dv/dt := g - a*v^2, with a :=0.115.

Complete and activate the following command.

```
> fsolve( 0= ... , v);
>
```

Place your answer below.

Answer:

Estimate the time it takes a large raindrop to fall to the ground from 3000 feet by assuming that the velocity is the constant terminal velocity the during the full duration of the fall. How does this answer compare with the time it takes a small raindrop to fall 3000 feet, as estimated in Section 3?
Enter your computations below.

Answer:

□ ADDENDUM: Maple Explanations.

In this section we give further explanations of some of the Maple commands used in the Worksheet. The information presented here will not be of crucial importance for completing or understanding the laboratory--if it was, it would be presented in the Preliminaries or Project sections. However, it should increase your understanding of the statements you are asked to deal with, and--if a problem arises--may help you in ferreting out the cause and solution.

¤ Functions that Remember Values.

The Maple "proc" construction used to define the recursive function v has a built-in "options remember" function which builds a remember table to store the computed values which makes computing recursive functions much more efficient. For example, in defining Fibonacci numbers, 1, 1, 2, 3, 5, 8, ... , proceed as follows:

```
> fib := proc(k)                          # This format uses the "option remember"
>    fib(k):= fib(k-1) + fib(k-2);   # which establishes a remember table.
>    end;
> fib(1) := 1;                            # Writing these initial values here puts them
> fib(2) := 1;                            # in the remember table. Activate this region.
```

The remember table of a procedure is accessible as the fourth operand of the procedure structure. Activate the following region and see the two entries.

```
> op(4, eval(fib));
```

Now, activate the next two lines.

```
> fib(4);
> op(4, eval(fib));
```

Maple provides the utility function "forget" to remove one or all entries from the remember table. Activate the region.

```
> readlib(forget): # Load the utility function.
> forget(fib,4):      # Forget the result of fib(4);
> op( 4, eval(fib)); # Update the remember table.
```

To empty the remember table of fib , enter the next two lines.

```
> forget(fib);
> op( 4, eval(fib));
>
```

This should give you an idea of how the option remember works.

The tvData and tv2Data commands also have the "options remember" option. Thus, if you calculate tvData(5) and then tvData(20), the remember table will store the entire two data sets, so that if you were to call on them again--say to plot the data set--the tvData function simply goes back to its "memory table" and pulls out the stored values. This is fast and efficient.

The disadvantage of this remember procedure is that tvData is now a "pack rat"--it will store every value that you ask it to compute! If you are not careful, this could cause the computer to run out of memory and crash. But don't reach for the "forget" command too quickly! If you also wish to retain some of the computed data sets, then you must first save them under another name before the forget statement is executed, e.g., with a statement such as
tvData40 := tvData(40). □

LAB 7: Derivatives and Zeros

□ NAMES:

□ PURPOSE
The purpose of this laboratory experiment is to explore, graphically and numerically, the relation between the derivative of a polynomial and the polynomial itself. In particular, we will study the relationship between zeros of the derivative and various features of the original function.

□ PREPARATION
Study Section 4.1 of Calculus: Modeling and Application. You should also read this printed copy of the lab notebook prior to your laboratory session.

□ LABORATORY REPORT -- General Instructions.
Note: Follow your instructor's instructions if they differ from those given here.

You are to write a self-contained, well-organized exposition and analysis of the issues raised in this laboratory. In particular, it should incorporate the answers to the important questions raised during the lab session.

Two submissions will be made for this report. Comments made on the first submission will help you prepare the final submission, which is the copy that will receive a grade. Moreover, the grade will be based solely on the written report--the instructor will look at other parts of the worksheet only in unusual situations. You should therefore consider the entries you make in the Preliminaries and Project sections to be notes to help when later writing your report.

More specific instructions on the content of the report will be given at the end of the worksheet.

□ Maple procedures: Please activate the following Maple command.

```
> with(lab7);
> # If an Error is an output of the above line, see your lab instructor.
```

Lab 7 Procedures provide us with two new Maple commands, DerivativeZeros and FindZeros. Both will be useful in generating pictures when examining the zeros of the derivative function. The commands in this package merely call the major command FunctionPlot.

□ PRELIMINARIES
¤ Derivatives in Maple.

Consider the following polynomial function:

p(t) = 3*t^4 - 16*t^3 + 6*t^2 + 24*t + 1

Enter a Maple function statement to define p.

```
> p := t -> ???;
```

We now wish to compute the derivative of p(t). Compute your answers on paper, then activate the next line.

```
> Ans(1);
>
```

• Maple Tip. Maple does not use the "prime" notation for derivative. In fact, given a function p (defined with the arrow notation), Maple denotes the derivative by D(p). Enter the expression below:

```
> D(p);
>
```

• Maple Tip. If the output to the above was "D(p)", then you need to go back and Enter the command defining p above. The D command is used with functions defined with the arrow notation. The output, D(p) is also a function defined with the arrow notation.

• Optional. As further practice,
(1) Enter the expression g(x) = x^2 + 3*exp(2*x) into Maple as a function using the arrow notation.
(2) Compute the derivative of g(x) with paper and pencil, and
(3) Compute the derivative using Maple and compare with your paper-and-pencil result.
Enter your computations below.

• Maple Tip. Maple, like most of us, distinguishes between functions and expressions. Mathematicians would say, the function f defined by the equation f(x)=x^2, or the rule that assigns to each x the expression x^2. Maple's notation for a function f := x -> x^2 is very much like the later. Then f(x) is the expression x^2; note, expression as opposed to a function! Activate the following two lines.

```
> f := x -> x^2;
> f(x);
```

Maple also gives names to expressions as follows: h := x^3 . Compare with arrow notation.

```
> h := x^3;
```

Maple computes derivatives of "expressions" with the "diff" command, but the variable must be specified. Here are some examples. Enter the following.

```
> diff(h, x);  # This assumes that the command line above has been activated.
> diff(x^3, x);   # Asks for derivative of x^3 with respect to x
```

See what happen if you leave off the variable.

```
> diff(x^3);
> diff(2*exp(t/3) - 2*t^3 + t^5, t);
> diff( t^2, x);              # Note change of variable.  Is the output correct?
>
```

• Optional. Ask Maple to compute the derivative of

x^5 -2x^3 +exp(2*x)*cos(x^2).

Enter your calculations below.

Hopefully you are curious about the rules Maple applies to compute these complicated derivatives. You will learn them all by the end of the semester.

□ PROJECT

¤ 1. Zeros by Zooming.

Recall the polynomial function p(t) defined in Preliminaries:

p(t) = 3*t^4 - 16*t^3 + 6*t^2 + 24*t + 1

Graph the function p(t) over some interval on the t-axis. Pick your interval so that all the zeros for p(t) are clearly visible. (Remember: this requires the plot(...) command.) Enter the plot statements below:

• Maple Tip. Try specifying just the x-range first and note what Maple picks for the y-range. After viewing a plot, exit it with either the File, Exit (mouse) or Alt+F4 (keyboard). The alternative is to click on the worksheet and hide the plot, but it is sitting back and eating up memory. You can find it with Alt+Tab.

```
> plot(p, ??? .. ???);
```

Now specify a y-range.

```
> plot(p, ??? .. ???, ??? .. ???);
>
```

Approximate all the zeros of this function by zooming in, i.e., by appropriately reducing the size of the t-interval until the zeros have been determined to within three decimal place accuracy. This means that the length of the bounding t-interval is no greater than .0005 (i.e., the error of approximating the root by either t-interval endpoint is no greater than .0005). Do this for each root, leaving below the last plot statement used in each case. Complete and Enter your plot statements below.

```
> plot(p(t), t = ??? .. ???, ??? .. ???); # Choose shorter & shorter intervals
> plot(p(t), t = ??? .. ???, ??? .. ???);
> ??? etc...
>
```

Summary. How many roots did you find? List them below, to three decimal place accuracy. List roots below:

roots := [??? , ??? , ??? , ???];

¤ 2. More Zeros by Zooming.

• Maple Tip. As seen above, the "plot" command can be used with either functions or expressions, but there is a difference in how you specify the x ranges.
 1. Functions, f := x -> x^2, use plot(f), plot(f,a..b), or plot(f, a..b, c..d), where x is not specified. If the x-range, a..b, is not given, the default -10..10 is used.
 2. Expressions, f(x), use plot(f(x), x=a..b), or plot(f(x), x=a..b, y=c..d).

```
> f := x->x^2;
> plot(f);
> plot(f, -1..1);
> plot(f(x), x= -1..1);
> plot(x^2, x=-1..1, y=-1..2);
```

We continue working with the polynomial p(t) from the previous section.

Graph the derivative function D(p) over some interval on the t-axis. Pick your interval so that all the zeros for D(p) are clearly visible. Enter your plot statement below:

```
> plot(D(p), ???, ???);
> # Remember to Exit your plots before continuing.
```

Approximate all the zeros of this derivative function by zooming in, i.e., by appropriately reducing the size of the t-interval until the zeros have been determined to within three decimal place accuracy. Do this for each root, leaving below the last plot statement used in each case. Complete and Enter your plot statements below.

```
> plot(D(p), ??? .. ??? ); # Choose shorter & shorter intervals E.
> ??? etc ...
```

Summary. How many roots did you find? List them below, to three decimal place accuracy.

```
> derivativeRoots := [ ??? , ??? , ??? ];
>
```

¤ 3. The Significance of a Zero for the Derivative.

We continue working with the polynomial p(t) from the previous sections.

Our goal in this section is to see what happens to a polynomial p at the points where its derivative D(p) is zero.

We will examine the polynomial p and its derivative D(p) by plotting both together in one graph. The Maple commanded needed for this task is the plots[display] command.

Let pPlot be the plot of the expression p(t) over the t-interval from -1.5 to 4.7 (these values were chosen simply to include all of the zeros you discovered in the previous sections). Similarly, let pDerPlot be the plot of the derivative function D(p) over the same t-interval. We place both graphs together on the same coordinate axes by combining them in a plots[display] command. That's all there is to it. We give the Maple commands below. Enter the commands given below.

• Maple Tip. The first statement shown ends in a colon : which suppresses any output from the commands. When a name, like pPlot, has been defined to be (:=) a plot command that has been terminated by a semi-colons (;), the output is a collection of technical Maple graphics commands. Try it. Obtain a plot by ending the command with :"; . The colon causes the plot structure to be put into the computer memory, the quote " tells Maple to use the last command, and the semicolon brings up the plot. Both plots will be in memory and the plots[display] command puts them into the same plot.

```
> pPlot := plot(p, -1.5 .. 4.7, -85 .. 40, color = black):
> pDerPlot := plot(D(p), -1.5 .. 4.7, color = blue):";
> plots[display]({ pPlot, pDerPlot });
>
```

For your convenience we have automated this process with a special command: DerivativeZeros. The statements given above form the core of DerivativeZeros, although additional features have been added. We apply this command to the polynomial p. Enter the command below.

```
> DerivativeZeros(p, -1.5 .. 4.7, -85 .. 40);
>
```

You should see some dots on the t-axis. These are the zeros of the derivative. For your convenience, they are further connected by vertical lines to the corresponding points on the polynomial graph. Remember: the purpose of this section of the lab is to see what happens to a polynomial p at the zeros of its derivative.

Study the graph of p and its derivative that you generated above. Do you see any special property at those points on the graph of p for which the derivative D(p) is 0? Before you either give up without a clue, or--at the opposite extreme--jump to a hasty conclusion, try some further examples. Define your own polynomials--call them p1(t), p2(t). Then run them through DerivativeZeros, and analyze what you get. Look for some special property of the polynomial that is found (most of time) at zeros of the derivative! Complete and Enter the commands below.

• Maple Tips. When looking at a new polynomial you'll have to experiment with a few different pairs of t limits before you achieve the best results.

```
> p1 := t -> ??? ; tmin := ??? ; tmax := ??? ; ymin := ??? ; ymax := ??? ;
> DerivativeZeros(p1, tmin .. tmax, ymin .. ymax);

> p2 := t -> ??? ; tmin := ??? ; tmax := ??? ; ymin := ??? ; ymax := ??? ;
> DerivativeZeros(p2, tmin .. tmax, ymin .. ymax);
>
```

The Property. What is the special property that you've identified at (most of) the points where the derivative is zero? Call any point on the graph of a polynomial an M-point if it has this special property. Describe below the special property that defines an M-point.

Answer:

The Precise Relationship. Decide if either of the following two statements is always true for any polynomial p : (1) If t is a zero of the derivative, then (t, p(t)) is an M-point; (2) If (t, p(t)) is an M-point, then t is a zero of the derivative. Place your conjectures below.

¤ Conjectures:

Test your conjectures with q(t)=2*t^3+1. Do the results of this test support your claims? Enter the following command, then place your answers and justifications below.

```
> q := t -> 2*t^3+1:
> DerivativeZeros( q, -1 .. 1, -2 .. 2 );
>
```

Answer:

Justification. Give an intuitive geometric justification for your conjecture about M-points and zeros of the derivative. Base your argument on the interpretation of the derivative as the slope of a tangent line. Place your answer below (two or three lines should suffice).

Answer:

¤ 4. The Relationship between Polynomial Zeros and Derivative Zeros.

We continue working with the polynomial p from the previous sections.

Our goal in this section is to see if there any relationship between the zeros of a polynomial and the zeros of its derivative.

Studying this question will be aided by a special command: FindZeros. Here is an example. Enter the following command.

```
> FindZeros(p, -1.5 .. 4.7, -85 .. 40);
>
```

Notice that the zeros of both the polynomial and its derivative are shown: the polynomial zeros are black dots, while the derivative zeros are blue dots.

What do you notice about the location of the zeros of p in relation to the zeros of its derivative D(p)? Can you formulate a general rule about the location of the zeros of a polynomial and the zeros of its derivative? (And remember: Rules can start in different ways, e.g., "Given two zeros of p , then," or "Given two zeros of D(p), then" The choice you make will be important.)

Before you either give up or jump to a hasty conclusion, try some further examples. Define your own polynomials--p1, p2,--and then test them in FindZeros. Complete and enter the following commands.

```
> p1 := t -> ??? ;  tmin = ??? ;tmax = ??? ;  ymin := ??? ;  ymax := ??? ;
> FindZeros( p1, xmin .. xmax, ymin .. ymax );

> p2 := t -> ??? ;  tmin = ??? ;tmax = ??? ;  ymin := ??? ;  ymax := ??? ;
> FindZeros( p2, xmin .. xmax, ymin .. ymax );
```

The Relationship. Describe the relationship(s) you have discovered between the location of the zeros of a polynomial and the zeros of its derivative. (State it--or them--precisely.) Place your conjecture(s) below.

Conjectures:

Are you sure of your conjectures? Torture test them on the polynomial q1(t)=p(t)+25 using FindZeros. Do the results of this test support the claims you've made above? If not, then reformulate the claim and test it with further examples. Enter your commands and answers below:

```
> FindZeros(t -> p(t)+25, -1.5 .. 4.7, -70 .. 50);
>
```

Answer:

Justification. Give an intuitive geometric justification for your conjectured relationship between zeros of the polynomial and zeros of the derivative. (Hint: The slickest argument justifies the current result on the basis of the previous section, i.e., on the relationship between M-points and zeros of the derivative!) Place your answer below.

Answer:

Remark: If your conjectures in this section are formulated correctly, then you have discovered a famous calculus result known as Rolle's Theorem (as least the version for polynomials).

¤ 5. What About Non-Polynomials? (Optional--Ask your instructor.)

The conjectures you made about the zeros of polynomials and their derivatives in the previous two sections carry over to a more general class of functions. However, in some instances there are changes that need to be made in the conjectures when applied to non-polynomials.

Derivative Zeros and M-points. Consider the conjecture you made relating zeros of the derivative to M-points on the polynomial. Test your conjecture for more general functions by generating examples using DerivativeZeros Restrict your attention to intervals on which the function in question is differentiable. Interesting candidates you might consider (among many others) are: cos(x)-exp(x)*sin(2*x) and (x^3-x^2-3*x+2)/(x^2-9). Enter your examples below.

```
> g := x -> ??? ;  xmin := ??? ;  xmax := ??? ;  ymin := ??? ;  ymax := ??? ;
> DerivativeZeros(g, xmin .. xmax, ymin .. ymax);
>
```

In view of these examples, do you think your conjecture concerning derivative zeros and M-points is valid for general differentiable functions on an interval? Place your answer below.

Answer:

Derivative Zeros and Polynomial Zeros. Consider the conjecture you made relating zeros of the polynomial to zeros of the derivative. Test your conjecture for more general functions by generating examples using FindZeros Restrict your attention to intervals on which the function in question is differentiable. You might again consider the functions: cos(x)-exp(x)*sin(2*x) and (x^3-x^2-3*x+2)/(x^2-9). Enter your examples below.

```
> g := x -> ??? ;  xmin := ??? ;  xmax := ??? ;  ymin := ??? ;  ymax := ??? ;
> FindZeros(g, xmin .. xmax, ymin .. ymax);
>
```

In view of these examples, do you think your conjecture concerning polynomial and derivative zeros is valid for general differentiable functions on an interval? Place your answer below.

Answer:

□ LABORATORY REPORT--Specific Instructions.
Note: Follow your instructor's instructions if they differ from those given here.

As usual, your observations are important--but your explanations of those observations are equally important! Your report should be a coherent, self-contained, and well-organized account of the major ideas in this laboratory exploration. You may assume a reader who has some calculus knowledge, but you cannot assume a reader who is familiar with the earlier parts of the Worksheet, i.e., do not assume the reader was in the lab with you!

Here are some further suggestions for your report. They are designed to get you started and stimulate your thinking--they are not designed to be a straight-jacket. If you have good reasons for wishing to organize your report in a different manner, feel free to do so.

The report should focus on the most important results of the lab. For example, the preliminary material on computing derivatives in Maple need not be mentioned, and you do not need to spend a great deal of time in describing the zooming method for finding roots of polynomials--a paragraph at most will do.

The report should concentrate on the material in Sections 3, 4, and 5. In particular, you should clearly state the conjectures you made in Sections 3 and 4, and give evidence (e.g., copy a representative graph or two) for your claims. It is also important to give clear geometric justifications for the claims, as we requested during the lab session.

There were a number of plausible but false conjectures that could have been made in each of Sections 3 and 4. A really top notch report will give these false conjectures and then show why they are false with, say, graphs where the conjectures visibly fail.

If you completed Section 5--extending your claims from Sections 3 and 4 to more general functions than just polynomials--the results could be discussed as the last part of your report, or they could be blended in with your discussions of the two individual claims. A superior report might consider what could go wrong with the conjectures for more general functions. This means considering what could happen when a function is not differentiable over the interval in question. (This would be a non-trivial task, however, since we did not deal with any examples of this nature.)

Place your report below.
□

LAB 8: Newton's Method

☐ NAMES:

☐ PURPOSE
The purposes of this lab are to study good and bad features of Newton's Method for solving nonlinear equations and to compare this method with the zooming method used in a previous lab.

☐ PREPARATION
Study Section 4.3 of Calculus: Modeling and Application, review your work in the Derivatives and Zeros lab on approximating zeros of functions by "zooming," and read the instructions in this worksheet prior to the lab session.

☐ LABORATORY REPORT INSTRUCTIONS
Note: Follow your instructor's instructions if they differ from those given here.

For this project you are not required to submit a formal report. Rather you are asked to submit a copy of this Worksheet with the questions answered in the indicated locations. Only one submission will be made. Write in complete sentences and be sure the sentences connect together in coherent ways.

☐ Maple Procedures. Please Enter the following line.

```
> with(lab8);
> # If an Error is an output of the above line, see your lab instructor.
```

The procedures just loaded provide us with a number of new Maple commands, including NewtonGraphs and SpeedAnalyzer. NewtonGraphs will generate pictures of the iterative steps in Newton's Method for approximating roots of equations. SpeedAnalyzer helps us study the speed of convergence of Newton's Method. The major routine of the package is Newton, although we will not use it directly--we access it indirectly since it is called by NewtonGraphs.

☐ PRELIMINARIES
¤ 1. Approximating to n Decimal Place Accuracy.

Suppose x0 is an approximation for a number x. The error of the approximation is given by

error = x - x0

Complete the third and fourth entries. We say

x0 approximates x to 2 decimal place accuracy if | error | < .005
x0 approximates x to 3 decimal place accuracy if | error | < .0005
x0 approximates x to 4 decimal place accuracy if | error | < ???
x0 approximates x to 5 decimal place accuracy if | error | < ??? And so on.

Determine the decimal place accuracy of the following approximations.

x0 = 2.34335, x = 2.34678 Answer:
x0 = 2.34115, x = 2.34678 Answer:
x0 = 2.99996, x = 3.00000 Answer:

Are you "surprised" by any of your answers? Check them against our answers:

```
> Answer(0);
>
```

Now consider the function dpa(x,x0), this will compute the number you obtained by hand automatically.

To see what dpa(x,x0) means, evaluate it at the three x, x0 pairs you looked at above. Complete, then enter the following command.

```
> [ dpa(2.34335, 2.34678),  dpa( ??? ,  ??? ),  dpa( ??? ,  ??? ) ];
```

In view of these numbers, what do you think the function dpa(x,x0) calculates? What do you think the three letters in the name dpa stand for? Make a conjecture, then Enter the next line.

```
> Answer(1);
>
```

We will use the function dpa(x,x0) later in the worksheet when measuring the speed of convergence of Newton's Method to zeros of functions.

¤ 2. Newton's Method--the Derivation.

Newton's Method is a procedure for approximating the zeros (also called roots) of a function y=f(t). For the function y=f(x) graphed below, we see that a root exists between 1.15 and 1.2--in this section we will explain the geometric procedure used in Newton's Method for generating a sequence of better and better approximations to this root. We begin with t=t(0), an initial approximation for the root (in the example below we start with t=t(0)=1.1).

```
> Graphic(1);
```

We now want to generate an improved approximation, t=t(1), for the root near 1.2. To do this we start by moving vertically from (t(0), 0) to the corresponding point on the graph of y=f(t).

```
> Graphic(2);
```

Determine the coordinates of the point you reached on the graph of y=f(t). (Write your coordinates for the general case y=f(t), starting with t=t(0), not for the specific example.) Write your answer on paper, then check the following answer.

```
> Answer(2);
```

Now "slide along" the tangent line at this point until you cross the t-axis. In our example the crossing occurs at approximately 1.22, as seen below.

```
> Graphic(3);
>
```

The t-value at this intersection is the next approximation, t=t(1), to our root (in our example, t(1) is approximately 1.22). In order to turn this geometry into a useful computational procedure we need to determine a formula for t(1). To do so, we use the slope of the tangent line. Express this slope in two different ways, i.e., with a derivative and without a derivative. Write your answers on paper, then enter the following command.

```
> Answer(3);
```

Since a line has only one slope, your two expressions for the slope must be equal. Thus set them equal to each other and solve for t(1). This is the expression for t(1) that we desire. Write your answer on paper, then check.

```
> Answer(4);
>
```

We can use the same process to go from t(1) to a t(2), then from t(2) to a t(3), etc., generating a sequence of approximations to the desired root. Give an equation for t(k+1) in terms of t(k). Write your answer on paper, then check.

```
> Answer(5);
>
```

This is the Maple code for Newton's Method that you will use during the Project.

The procedure t represents the recursive definition of Newton's method. The A is a list showing both the iteration level and the iteration result of Newton's method.

```
> t := proc(k)  t(k) := t(k-1)-f(t(k-1))/D(f)(t(k-1)) end;
> A := seq([ k, t(k) ], k=0..n);
>
```

□ PROJECT.

¤ 1. An Example of Newton's Method.

We consider the polynomial function that was studied in a previous laboratory:

f(t) = 3*t^4 - 16*t^3 + 6*t^2 + 24*t + 1

Enter a Maple function statement to define f(t) below.

```
> f := t -> ??? ;
```

Graph the function f over some interval on the t-axis. Pick your interval so that all the zeros for f are clearly visible (you will probably need to experiment with the interval values). Enter your plot statement below.

```
> plot( ???, ??? );
>
```

We will approximate each of the four roots of the polynomial f by the use of Newton's Method.

Test run on Maple commands below.

1) Set n := 2 ; later you will put it at 10 .

2) Set t(0) := 3 and activate the region. The output contains fractions, but no decimal numbers and this is bad. Computations with fractions is memory intensive and if n := 10, your machine would have "crashed". If output contains the symbol f, then Enter f again.

3) Change 3 to 3., a 3 followed by a decimal point "." and activate region. If output contains decimals, this is what you want.

• Real Thing. Decide on an initial guess t(0) for one of the four roots of f--choose this value by eyeballing the plot of f above. Then Enter this value for t(0), but make sure it has a decimal point . in it. Say "t(0) := 3." . Check again with n :=2 and then change n to 10. Are the results expected?

```
> Df := D(f):
> t := proc(k) t(k) := t(k-1)- f(t(k-1))/Df(t(k-1)) ; end:
> t(0) :=3.:     # Include a decimal point, even if it looks like 2.0 or 2. .
> n := 2:
> A := seq( [k, t(k) ], k=0..n);
>
```

• Maple tips:

1. For t(0), do not use 0, or even 0.0 , since Maple considers both to be an integer and computes fractions! Use 0.1 or 0.01, etc.

2. A way to avoid these problems is to use

t(k) := evalf(t(k-1)-f(t(k-1))/Df(t(k-1)))

in the 2nd line. The evalf command evaluates the objective as a decimal.

3. When defining a recursive function as we are here, the initial conditions (t(0):=???) must be identified after the function is defined. This allows Maple to easily generate a table in memory for the function that speeds up evaluation.

Record the value of the root just found, and then approximate the remaining three roots in the same fashion. (For each remaining root, simply change t(0) to a rough approximation of the root, with a decimal, and then re-execute the Newton's Method code above.) Record the values of the four roots that you find as r[1], r[2], r[3], and r[4] (in increasing order). Type and Enter your values for the four roots below (in increasing order).

```
> r[1] := ??? ; r[2] := ??? ;
> r[3] := ??? ;
> r[4] := ??? ;
```

Which method do you prefer: the "zooming in" method of the previous lab or Newton's Method? Place your answer below.
Answer:

¤ 2. Newton's Method--A Visual Demonstration.

We have constructed a sophisticated routine that generates pictures for the iterative procedure used in Newton's Method. You will be able to take any reasonable function, make a guess for a root, and then sit back and watch Maple iterate its way to an approximation for the root.

Our graphics routine is invoked by the command NewtonGraphs. Execute the information command ?NewtonGraphs to learn how NewtonGraphs works. Enter your information command below.

The explanation produced by your information command may look daunting, but using NewtonGraphs is not difficult. To run it, define the quantities shown in the parameter list:

f, the function whose roots are to be approximated,
xmin and xmax, the initial interval in which to look for a root,
x0, the initial approximation to a root in the interval from xmin to xmax (if no value is specified, then x0 defaults to (xmin + xmax)/2).

You should also decide whether or not you want to zoom in on the root in your pictures. If so, then include the option

"zooming=true"

at the end of the parameter list. Once these quantities are given, NewtonGraphs will give a good show. It is advisable to take initial runs without zooming. Once you see that your iteration is relatively well-behaved, then a run with "zooming=true" makes sense. Below is an example using the current function f. Enter this and examine the pictures. Enter the following Maple commands (after reading the Warning).

• Maple Warning!! Be sure you have Saved your work before executing this command! A lot of graphic images are going to be generated--these could cause Maple to unexpectedly abort if you are low on available memory.

```
> NewtonGraphs(f, -1.5 .. 4.7,start=3.2, zooming =true):
>
```

From the output does it appear that our successive approximations are heading to a root for the function? You can tell by looking at the function values at the successive values of t--are these values heading toward zero? Place your answer below.
Answer:

Now rerun the example, this time with zooming=true included as an option. To do this, you have only to go back to the NewtonGraphs command you executed above and add in the statement zooming=true at the end of the parameter list, i.e., ...=3.2, zooming=true) Then reEnter the altered command. Edit the previous NewtonGraphs command to include zooming, then Enter it.

Describe how the local linearity of the function is speeding up the convergence to the root. (A well behaved function is called "locally linear" because when you examine any small portion of its graph under a microscope, you get what appears to be a straight line. This line must coincide with the tangent lines to points in that region.) Place your answer below.
Answer:

Let's make one more pass with this same example, but this time without pictures. In other words, we are going to use NewtonGraphs as a Newton's Method evaluation routine, not as a graphing routine. To do this, we merely add the option plotting=false into the NewtonGraphs command. Enter your commands below.

```
> NewtonGraphs( ??? );
>
```

The result from NewtonGraphs when using plotting=false is a little complex. The first number is the approximated root and the second number is the number of iterations needed to get that close. More precision can be applied with the minDig option. Set minDig=20 and maxIter=20 and see what happens.

How many iterations were necessary before the desired accuracy of convergence was reached? Place your answer below.
Answer:

You've been investigating a case where Newton's Method works perfectly. In the next section you will use NewtonGraphs to investigate the limitations of Newton's Method.

¤ 3. Some Limitations of Newton's Method.

In this section you will run Newton's Method for f and the following five initial values: Enter the following definitions.

```
> s[1]:=0.9; s[2]:=1.1; s[3]:=3.56; s[4]:=3.57; s[5]:=1.0;
>
```

Indicate below to which roots--r(1), r(2), r(3), or r(4)--you "expect" convergence when starting with each of the indicated values. (In some cases you might list two possibilities.) Your "expectations" will probably be based on which root r(j) each of the guesses s(k) is closest to. Place your answers in the space below.

Expected Convergence Results: We might expect to have convergence as follows:

s[1]=0.90 should go to r[???]
s[2]=1.10 should go to r[???]
s[3]=3.56 should go to r[???]
s[4]=3.57 should go to r[???]
s[5]=1.00 should go to r[???]

Does Newton's Method converge to the expected root for each of these initial values? For each of the initial values s[k] run NewtonGraphs (without plots) to see if you get convergence to the expected root. Also look at the t[k] values and their function values. Anything funny happening here? Record the convergence results (i.e., s[1] goes to r[]) below. (For your convenience we have included the necessary command) Enter the following code for each s[k]. Record your results below.

```
> x0 := s[4];
> NewtonGraphs(f, -10 .. 10, start= x0, plotting=false);
>
```

Actual Convergence Results: We achieved convergence as follows:

s[1]=0.90 went to r[???], number of iterations = ???
s[2]=1.10 went to r[???], number of iterations = ???
s[3]=3.56 went to r[???], number of iterations = ???
s[4]=3.57 went to r[???], number of iterations = ???
s[5]=1.00 went to r[???], number of iterations = ???

Copy your Expected Convergence Results region and Paste it below. Did you get what you expected for each of the initial values? Place your answers (Yes or No) next to the expectations listed below.

In order to understand what is going on here, generate graphs for one or two iterations of Newton's Method for each of the five initial values. We wish to find an answer to the following question: What property of the tangent line at the initial point can cause Newton's Method to behave strangely? (For your convenience we have included the necessary commands) Enter the following code for each s[k].

```
> NewtonGraphs(f, -1.4 .. 4.7, start = 0.9 );
>
```

What property of the tangent line at the initial point can cause Newton's Method to behave strangely? Express this in terms of the derivative at the initial point. Place your answer below.
Answer:

How could we "test" potential starting values for Newton's Method to prevent these problems? Place your answer below.
Answer:

¤ 4. Computing the Exact Roots using Maple.
You will need f(t) from Section 1.

You should recall that we can find the roots of any 2nd degree polynomial algebraically. The formula for the roots in such a case is called the Quadratic Formula. Similarly, formulas exist for the roots of any 3rd or 4th degree polynomial, although they can be very complicated. In particular, since f(t) is a 4th degree polynomial, we can compute its roots exactly. Well, more truthfully, Maple can do it. Enter the following command:

```
> f := t -> 3*t^4-16*t^3+6*t^2+24*t+1;
> solutions := solve(f(t));
>
```

Maple will return the problem unevaluated if it does not see an obvious solution. Numerical methods can then be used to get any degree of accuracy desired. We want very good approximations, so let's ask for 20 accurate digits. Enter the following command:

```
> Digits := 20;
> Nsolutions := [ allvalues(solutions) ];
> for i to 4 do rt[i] := Nsolutions[i] od;
>
```

• Maple Tip. The command allvalues in this context will find all the roots, instead of just one. Digits sets the number of digits of accuracy desired. Any number can be used.

We can now work with this list of numbers as we have done in the past. In particular, our 20 decimal place approximations to the four roots will be designated by rt[1], rt[2], rt[3], rt[4]. Since they are in increasing order, they correspond to the approximations r[1], r[2], r[3], r[4].

¤ 5. Analyzing the Speed of Convergence of Newton's Method.

You will need f(t) and [r[1], r[2], r[3], r[4]] from Section 1, as well as roots from Section 4.

For each of the four roots of our polynomial f(t) we intend to re-run Newton's Method. However, this time we will make use of our knowledge of the exact roots (as obtained in the previous section) to study the speed of convergence, i.e., how fast the Newton approximations approach a root. We measure the speed of convergence by computing the decimal place accuracy of each iteration x[k] as an approximation to a root x, i.e., the quantity

dpa(x, x[k]),

where dpa is the function defined in the Preliminaries section of this lab.

For your convenience, in the following command line are four initial starting values for Newton's Method, t[1], t[2], t[3], and t[4]. These starting values will yield Newton iterations that converge to the roots rt[1], rt[2], rt[3], and rt[4] respectively. These will provide good starting points for the decimal place accuracy study. Enter the following line.

```
> t[1]:=-1.6; t[2]:=.67; t[3]:=1.4; t[4]:=5.5;
>
```

The computations for this will be done by another Maple command: SpeedAnalyzer. Find out what it does by executing the usual ? command. Enter your command here.

The best way to understand a routine is to have an example. Enter the following group of commands.

```
> f := t -> 3*t^4 - 16*t^3 + 6*t^2 + 24*t + 1;
> maxIter := 15:
> SpeedAnalyzer(f, rt[ 1 ], t[ 1 ], maxIter);
>
```

The table output has three columns. The first column is merely the counter k, the number of times Newton's Method has been used. The second column is x[k], the k-th approximation to the chosen root rt[...], computed to 20 digits. The third column is the decimal place accuracy of the k-th approximation -- "Exact" here means precise to at least 20 decimal places which is the maximum accuracy possible with the settings we use.

Run SpeedAnalyzer for all four pairs {rt[j], t[j]}, j = 1,2,3, and 4. Observe the shape of the dpa plot in each case! Do the points seem to lie on a straight line, i.e., is the plot of the decimal place accuracy verses iteration linear? Place your answer below.

Let's perform a mental analysis of the speed of convergence for the zooming method for finding roots -- we used this in the previous lab. At each stage in the zooming process you eyeball the picture and narrow your root search to a smaller subinterval. Let's assume you can make the subinterval about one-tenth the length of the interval (that seems like a reasonable estimate of what we can do with the eye). Then the error in our approximation of the root should decrease by about a division of 10 at each iteration. What does this mean for the decimal place accuracy of our approximations at each iteration? In particular, what should dpa plots look like for the zooming method? In terms of the general behavior of the dpa, which method do you prefer, Newton, or "zooming"? Place your answer below.

¤ 6. The Engineer's Dilemma.

The method of zooming, as you used in the previous lab, can be made to work automatically, i.e., without the need of "human decision" at each step. We saw one way to do this in a class demonstration: the method of sections. In this method we bracket a zero of the function, bisect the interval, pick out one of the subintervals that has a root, and continue the process with the new interval.

Given this information as background, imagine now that you are the engineer and you've been given the job to design circuitry for the Solve key on a new calculator, a calculator that will also have graphical and symbolic capabilities. The purpose of the Solve key is to provide the user with the solution of equations of the form f(t) = 0, where f is a function that the user enters as a symbolic formula. Your first task is to decide whether to use some version of the zooming process or to use Newton's Method.

If your calculator is not competitive on the market, your company may fold -- and even if it doesn't, the engineers who made bad decisions will be laid off! To be competitive, your Solve procedure must be as automatic as possible, as foolproof as possible, as generally applicable as possible, as fast as possible, and as accurate as possible. Rate each of the two method with regard to these requirements, on an A, B, C, D, F scale. Give comments if needed to explain your choices. Place your answers below.

Ease of use -- how automatic is it?
Newton's Method:
Zooming Method:

Is it foolproof?
Newton's Method:
Zooming Method:

Any restrictions on its applicability?
Newton's Method:
Zooming Method:

How fast is it?
Newton's Method:
Zooming Method:

How accurate is it?
Newton's Method:
Zooming Method:
□

LAB 9: The Spread of Epidemics

□ NAMES:

□ PURPOSE
We investigate the "S-I-R" model for describing the spread of an epidemic through a population. The model is given as an initial value problem which involves three differential equations and three initial conditions. As in previous labs, we use Euler's Method to approximate the solution functions. The lab ends with a comparison of the model under consideration with the spread of the Hong Kong Flu epidemic through the population of New York City in 1968-69.

□ PREPARATION
Read Section 5.2 of Calculus: Modeling and Application. Here the differential equations for the S-I-R model will be discussed, and the equations implementing Euler's Method for graphical solutions will be derived.

An excellent source for additional development of the S-I-R model is Chapter 2 of Calculus using Mathematica, Keith D. Stroyan, Academic Press, 1993. This is the textbook for the calculus reform project at the University of Iowa.

□ LABORATORY REPORT INSTRUCTIONS.
Note: Follow your instructor's instructions if they differ from those given here.

For this project you are not required to submit a formal report. Rather you are asked to submit a copy of this Worksheet with the questions answered in the indicated locations. Only one submission will be made. Write in complete sentences and be sure the sentences connect together in coherent ways.

□ Maple Procedures. Be sure to activate the following command.

```
> with(lab9);
> # If an Error is an output of the above line, see your lab instructor.
```

The Lab 9 Procedures define the EpidemicModeler, a new command based on Euler's Method that will be the central tool of this lab, and a utility command called Zip that is a modification of the Maple command zip.

As an illustration, activate the following line.

```
> Zip( [1,2,3], [4,5,6]);
```

□ PRELIMINARIES.

¤ 1. Epidemics.

An epidemic is an unusually large short term outbreak of a disease. Human epidemics are often spread by contact with infectious people, though sometimes there are "vectors" such as mosquitoes or rats & fleas or mice & ticks involved in disease transmission. There are many kinds of contagious diseases such as smallpox, polio, measles and rubella ("German measles") that are easily spread through casual contact. Other diseases such as gonorrhea require more intimate contact. Another important difference between the first group of diseases and gonorrhea is that measles "confers immunity" to someone who recovers from it whereas gonorrhea does not. In other words, once you recover from rubella, you cannot catch it again. This feature offers the possibility of control through vaccination.

In this laboratory we will consider the S-I-R model for the spread of a disease like rubella where people are susceptible, then infectious, then finally recovered and immune. We will also assume that the only way to catch the disease is from another infected person.

¤ 2. The S-I-R model for epidemics.

Section 5.2 of Calculus: Modeling and Application develops the three differential equations and the three initial conditions used to model the spread of an epidemic. For convenience, we repeat them here:

```
dS/dt = - (C/N)*I(t)*S(t),                S(0) = N - I(0),
dI/dt =   (C/N)*I(t)*S(t) - L*I(t),       I(0) = Io,
dR/dt =   L*I(t),                         R(0) = 0.
```

• The variables S(t), I(t), and R(t) have the following meanings:
 S(t) is the number of susceptible individuals at time t,
 I(t) is the number of infected individuals at time t, and
 R(t) is the number of recovered (or deceased) individuals at time t.

• The two parameters L and C specify the nature of the disease:
 L (the recovery parameter) is the fraction of the infected group that will recover each day, and
 C (the contact parameter) is the number of contacts per day that an infected individual has which are sufficient to spread the disease.

• The initial values are as follows:
 N = the total population under consideration.
Since we will study New York City's Hong Kong Flu epidemic of 1968, N will be taken to be 7,900,000, the population of NYC in 1968.
 Io = the number of infected individuals at the beginning of the epidemic.
For the NYC example we will set Io = 10 (reflecting the fact that this was a "new" disease).

In the lab we approximate solutions to this initial value problem by using Euler's Method. Focusing, at least initially, on the Hong Kong flu, we use empirically determined values of C and L for this disease:

C = 0.47 and L = 0.34 (Hong Kong flu parameters.)

We model the epidemic over a time period T, which we initially set at 126 days (18 weeks), using a time step of a half-day, i.e., Deltat=0.5 days Then, starting with initial values S(0), I(0), and R(0), the values of these quantities at the time t(k)=k*Deltat will be denoted by S(k), I(k), and R(k).

• Given this set-up, the two primary goals of the lab are as follows.

Determine how the values of C and L affect the nature of the epidemic, at least as predicted by the S-I-R model. This will allow us to predict how epidemics differ from disease to disease, and will have public policy consequences on such issues as inoculation programs.

Determine how well the predictions of the S-I-R model match actual data from NYC's Hong Kong flu epidemic of 1968.

¤ 3. Deriving the "Updating Equations" for Euler's Method.

Given values for S(k-1), I(k-1), and R(k-1) , we wish to derive equations for the next set of values S(k), I(k), and R(k). To do this we will use Euler's Method, the central idea being the replacement of derivatives with difference quotients. At the (k-1)-th step define

DeltaS = S(k)-S(k-1), DeltaI = I(k)-I(k-1), and DeltaR = R(k)-R(k-1).

Replacing derivatives with difference quotients in the three differential equations in Section 2 will yield

(S(k) - S(k-1))/Deltat = -(C/N)*I(k-1)*S(k-1),
(I(k) - I(k-1))/Deltat = (C/N)*I(k-1)*S(k-1) - L*I(k-1),
(R(k) - R(k-1))/Deltat = L*I(k-1).

Solve these three equations for S(k), I(k), and R(k).

Compute your answers on paper, then activate the following line.

```
> Ans(1);
>
```

These equations, which tell us how to go from the (k-1)-th step to the k-th step represent Euler's Method and will be implemented in the next section.

¤ 4. Maple Generation of the S(t), I(t), and R(t) Graphs.

We want to see how our epidemic model predictions vary with different values of C and L. For that reason we start off with Maple commands (below) that define these values. The remaining commands define the recursive functions S, I, and R, set their initial conditions, compute the subsequent values of these quantities, and then plot the results as functions of time.

Enter the following commands and examine the output.

• Maple Comment. Maple reserves the symbol I for the imaginary number (-1)^.5 so for the Maple code we have changed S(k), I(k), and R(k) to the equally convenient S(k), In(k), and R(k).

```
> C := 0.47:   L := 0.34:   N :=7900000.:
>  tMax :=126.:   Deltat :=.5:
> kMax := floor(tMax/Deltat):
> t := k -> k*Deltat:
>
> S := proc(k)
>      S(k) :=  S(k-1)  - Deltat*(C/N)*In(k-1)*S(k-1);
>      end:
> S(0) := N-10:
>
> In := proc(k)
>     In(k) :=  In(k-1)  + Deltat*((C/N)*In(k-1)*S(k-1) - L*In(k-1));
>     end:
> In(0) := 10.:
>
> R := proc(k)
>     R(k) := R(k-1) + Deltat*L*In(k-1);
>     end:
> R(0) := 0:
>
> seq([S(k), In(k),R(k)], k=0..kMax):     # Computes and stores the values.
>
> plot1 := plot([seq([t(k),S(k)] , k=0..kMax)], color=COLOR1, thickness=2):
>
> plot2 := plot([seq([t(k),In(k)], k=0..kMax)], color=COLOR2,  thickness=2):
>
> plot3 := plot([seq([t(k),R(k)], k=0..kMax)], color=COLOR3,  thickness=2):
>
> plots[display]([ plot1, plot2, plot3 ]);
>
```

COLORi, i=1, 2, 3, has been defined in the Lab 9 Procedures so as to make the following exercise more meaningful.

• Problem. Eliminate the wrong choices in the following statements:
The susceptible population is the (red, blue, magenta) curve.
The infected population is the (red, blue, magenta) curve.
The recovered population is the (red, blue, magenta) curve.

You will carefully analyze these curves during this laboratory.

¤ 5. Maple Command: EpidemicModeler.

In order to make the Maple commands from the previous section easy to use, we have packaged them all into a new command: EpidemicModeler. To see what this does, specify allowable values for C and L (say 0.47 and .34 respectively, as used in the previous section) and then execute the following command.
Complete and Enter the following commands.

```
> C := ??? :
> L := ??? :
> EpidemicModeler(C, L );
>
```

Find out more about EpidemicModeler by executing the statement ?EpidemicModeler. Enter your command below.

As you can see, a number of options have been provided for EpidemicModeler. However, for the primary tasks of this lab, only two of the options will be required: plots and days:

• plots=infected will cause EpidemicModeler to only display the plot of the infected population. This is useful since it will enlarge the plot of the infected population.

• days=number will change the number of days over which you model the epidemic from the default value (currently 126 days) to any number you specify. This is useful if the chosen values of C and L produce an epidemic which is either very fast or very slow to take hold.

Try a run of EpidemicModeler with the option plots=infected included with the given constants, and then take a different value of C than you used above (say between 0.47 and 0.80). Enter your command below.

```
> C := .47:
> L := .34:
> plot1 := EpidemicModeler(C, L, plots=infected):";

> C := ??? : L := ??? :
> plot2 := EpidemicModeler(C, L, plots=infected):";
>
```

You've now made two runs of EpidemicModeler with the plots=infected option and have produced plot1 and plot2. How do the two epidemics you've modeled compare to each other, i.e., how do the two infection curves compare? To make such comparison easy to do, we use the plots[display] command which takes the output of any number of runs of EpidemicModeler and plots all the infected curves in one graph.

Enter the following command.

```
>   plots[display]([ plot1, plot2] );
>
```

EpidemicModeler and displaying several plots at once with plots[display] will be the primary tools used to study the S-I-R model.

□ PROJECT.

¤ 1. Varying the "contact parameter" C.

Remember: C--the contact parameter--is the number of contacts per day that an infected individual has which are sufficient to spread the disease.

Keep L fixed at 0.34 and experiment with the graph of I(t) by using different values of C between 0.47 and 0.80 in EpidemicModeler. As arranged for below, label your various plots as plot(1), plot(2), etc. Then combine them all into one graph using the plots[display] command and the plots=infected option. Complete and Enter the commands below.

```
> C := ??? ;  L := 0.34:
> plot(1)  := EpidemicModeler(C, L, plots=infected):
> C := ??? ;  L := 0.34:
> plot(2)  := EpidemicModeler(C, L, plots=infected):
> C := ??? ;  L := 0.34:
> plot(3)  := EpidemicModeler(C, L, plots=infected):
> C := ??? ;  L := 0.34:
> plot(4)  := EpidemicModeler(C, L, plots=infected):
> plots[display]([ plot(1), plot(2), plot(3), plot(4) ]);
>
```

What "geometric" effects did varying C have on the shape of the infected curve I(t)? Place your answer below.
Answer:

Should these effects have been expected given the meaning of C in the epidemic model as the number of contacts per day that an infected individual has which are sufficient to spread the disease?
Place your answer below.
Answer:

¤ 2. Varying the "recovery parameter" L .

Remember: L--the recovery parameter--is the fraction of the infected group that will recover each day. It is instructive to notice that 1/L is the average number of days that an individual remains infected with the disease.

Return C to 0.47, and experiment with the graph of I(t) by using different values of L between 0.1 and 0.55 in EpidemicModeler. As arranged for below, label your various plots as plot(1), plot(2), etc. Then combine them all into one graph using the plots[display] command. Complete and Enter the commands below.

```
> L := ???; C := 0.47:
> plot(1) := EpidemicModeler(C, L, plots=infected):
>
> L := ???; C := 0.47:
> plot(2) := EpidemicModeler(C, L, plots=infected):
>
> L := ???; C := 0.47:
> plot(3) := EpidemicModeler(C, L, plots=infected):
>
> L := ???; C := 0.47:
> plot(4) := EpidemicModeler(C, L, plots=infected):
>
> plots[display]([ plot(1), plot(2), plot(3), plot(4) ]);
>
```

What "geometric" effects did varying L have on the shape of the infected curve I(t)?

Place your answer below.
Answer:

Should these effects have been expected given the meaning of L in the epidemic model as the fraction of the infected group that will recover each day?

Place your answer below.
Answer:

As L approaches the end of the interval 0.1 <= L <= 0.55, a significant change occurs in the character of the graph of I(t). What is this change, and where does it occur? (Run EpidemicModeler a few more times if necessary.)
Enter your commands and answers below.

```
> L := ??? ;C := ???;
> EpidemicModeler(C, L, plots=infected);
>
```

Answer:

On the basis of your answer to the previous question, find a value Lo so that

if L>Lo, then an epidemic will not occur, while

if L<Lo, then an epidemic will occur.

If you change the value of C, how do you think the value of Lo will change?
Place your conjectures below.

Conjecture: L=???

* * * * * * *

• Theoretical Extra Credit Question. Use the differential equation

dI/dt = (C/N)*I(t)*S(t) -L*I(t)

to explain how you could have predicted in advance the value of L at which the character of the graph of I(t) changed. (Hint: For this "special" value of L, what do you suppose the value of dI/dt should be at t=0?)
Place your explanation and equations below.
Answer:

¤ 3. Examining all three curves.

Return L to its original value of 0.34 and C to its original value of 0.47. Graph all three curves, S(t), I(t), and R(t), by using EpidemicModeler.
Enter the following command.

```
> C := 0.47; L := 0.34;
> EpidemicModeler(C, L);
>
```

Do the three curves appear to be sensible, i.e., is the S-I-R model making predictions that are at least qualitatively reasonable? Carefully justify your answer by explaining in words how the epidemic predicted by the model proceeds from beginning to end. (We will see if the predictions are quantitatively reasonable in the next section.)
Place your answer below.
Answer:

¤ 4. Comparing the S-I-R model with the NYC Hong Kong flu epidemic of 1968.

We start this section with an excerpt from Keith D. Stroyan, Calculus using Mathematica, Scientific Projects and Mathematical Background, Project 3.

In the study of virulent diseases, one of the biggest problems facing investigators is the availability of sufficient data for analysis. In order to model a disease, we need to know the proportions of the population that are susceptible, immune, and removed. However, it is often very hard to determine these proportions because a large number of cases of a given disease go unreported (consider how many times you've had the flu, and out of those times, how many times you actually saw a doctor and so were "counted" as an infected). For this reason, certain diseases lend themselves to study because, in one way or another, they make their presence known to the medical community. For example, in the case of polio, the disease was severe enough to require medical attention, ensuring that there would be records of all cases. In the case of the Hong Kong flu, which you will be studying here, the data was derived from a slightly different source.

The data for the Hong Kong flu is derived from the medical reports of coroners in New York City. During the period of the epidemic, the course of the disease can be traced as a function of the excess of individuals who died due to influenza-related pneumonia in the New York City area. The term "excess" in the above sentence refers to the fact that, in New York City, you expect to get a certain number deaths due to influenza-related pneumonia. However, during the epidemic, there was a large increase in influenza-related deaths which could be attributed to the Hong Kong flu.

Once we have our data, the question becomes, how to interpret it? There are two factors that we must consider. First, not everyone who gets the Hong Kong flu is going to die as a result. And second, those who do die as a result probably will not do so right away, but rather will have the disease for an extended period of time before expiring. On the basis of these two facts, then, we can hypothesize that the number of excess deaths is proportional to the number of infected individuals from a couple of weeks before (the emphasis is ours).

Here is the New York City data on excess influenza-related pneumonia deaths during thirteen weeks of the Hong Kong flu epidemic. Enter the following commands which list and plot the excess deaths. For example, the 14 below says that during the first of the 13 weeks, there were 14 more deaths in New York City than the average number. The "days" command gives the number of days that have passed at the end of a given week.

```
> deaths := [14, 31, 50, 66, 156, 190, 156, 108, 68, 77, 33, 65, 24];
> days := [seq(7*i, i=1..13)];
> deathDays  := Zip(days, deaths);
> deathPlot :=plot(deathDays,  color=COLOR(RGB,0,0,1),  thickness=2,
>                      title=`Excess Flu Deaths, NYC, 1968`):";
>
```

It is important to realize that there is no significance to the starting time t=0 days in the "Excess Flu Deaths" plot. In particular, there is no reason to assume that t=0 in this plot corresponds to the "start" of the epidemic. Keep this in mind as we compare the S-I-R model against "Excess Flu Deaths."

Enter the following commands, which use the empirically determined values of b=0.47 and l=0.34 for influenza, to generate the S-I-R model for I(t), the infected curve. Enter the following commands.
EpidemicModler also accepts the option of title=`plot title`. See the next command.

```
> C := 0.47:  L := 0.34:
> modelPlot := EpidemicModeler(C, L, plots=infected, days=130,
>                                  title=`Predicted Infections`):";
>
```

In order to better compare these two graphs, we will:

• shift the "Excess Flu Deaths" plot over by a number shift (in days) so that the peaks of this plot and the "Predicted Infections" plot are almost equal, and
• multiply the excess flu deaths by a number, proportionality, so that the sizes of the (adjusted) "Excess Flu Deaths" plot and the "Predicted Infections" plot are approximately the same (remember our assumption that the excess flu deaths should be proportional to the intensity of the epidemic).

Assign values to shift and proportionality in the commands below, then Enter the region. Continue in this fashion until you have obtained as close a fit between the two curves as possible. (Hint: It is easy to estimate an initial value for shift by looking at the two curves. To obtain an initial value for proportionality, estimate the number that would be needed to multiply the largest "excess death" value to obtain the highest value on the S-I-R infection curve. If still in doubt, start with shift=25 and proportionality=1000.) Complete and Enter the following commands. Repeat until the best fit is obtained.

• Maple Tip. The shifting in the days := [7, 14, 21, ...] amounts to adding the "shift" to each entry of days. If x represents an entry in days, say 7, then x+shift is the desired shifted entry, 7+shift. The command map(x -> x+shift, days) shifts each entry in days.

```
> shift := ??? :
> days;  # If [7, 21,...] does not appear, reactivate the "days" command above.
> adjustedDays := map(x -> x+shift, days);
```

• Maple Tip. Multiplying each entry in deaths := [14, 31, 50, ...] by the number "proportionality" is done in an analogous way to form adjustedDeaths. Complete the following commands and activate.

```
> proportionality := ??? ;
> deaths:= [14, 31, 50, 66, 156, 190, 156, 108, 68, 77, 33, 65, 24];
> adjustedDeaths := map(x -> proportionality*x, deaths);
> adjustedDeathDays := Zip(adjustedDays, adjustedDeaths);
> adjustedDeathPlot := plot(adjustedDeathDays, color=COLOR(RGB,0,0,1),
>                           thickness=2):"; # Maximize plot and then exit it.
> plots[display]({modelPlot, adjustedDeathPlot} );
```

Were you able to get the two plots--the "Excess Flu Deaths" and the "Predicted Infections"--to match up in a satisfactory manner. On the basis of this comparison, do you think that the S-I-R model is valid for this type of an epidemic? Place your answer below.
Answer:

¤ 5. Levels of immunization and "herd immunity".

The purpose of a vaccination program against a specific disease is to develop immunity in a large enough proportion of a population that an epidemic cannot start. Having no epidemic is equivalent to the I(t) curve always going down. When such a condition is reached, the population is said to have achieved "herd immunity" to the disease. The question is, given the characteristics of a particular disease (the values of C and L), what is the level of immunization necessary to achieve herd immunity? We use the S-I-R model to predict the immunization levels for influenza, measles, and rubella.

Assume that in each case the population, N = 7900000, and the initial number of infected persons, i0 = 10.

• Influenza.

Given an influenza disease like the Hong Kong flu (C=0.47 and L=0.34), estimate the smallest fraction of the population that needs to be immune to guarantee that no epidemic will start. (This is a "Bisection" problem: If 0% of the population is immune, then an epidemic will start, while if 100% of the population is immune, then no epidemic will start. So the desired fraction lies between 0 and 1. Narrow these bounds by executing the following commands.) Complete and Enter the following commands.

```
> fractionOfPopulationWhichIsImmune := ???: # Fraction of initial population which IS imm
> une.
> C := .47 :
> L := .34 :
> population := 7900000:
> r0 := population * fractionOfPopulationWhichIsImmune:
> Days := 135:
> EpidemicModeler(C, L, initialImmunes=r0, plots=infected, days=Days);
>
```

What did you find as the fraction of the population that needs to be immune to influenza so as to guarantee that no epidemic occurs? Place your answer below.
Answer:

Suppose prior to 1968 the NYC Health Department had a vaccine for the Hong Kong flu that was (say) 95% effective, i.e., immunity would be developed in 95% of the individuals receiving the inoculation. What percentage of the population would the Health Department have had to inoculate prior to the epidemic in order to have prevented the epidemic? Would such a task have been possible? Place your answer below.
Answer:

• Measles.

The parameters for measles are approximately equal to b=2.00 and l=0.13. Determine the smallest fraction of the population that needs to be immune to guarantee that no epidemic will start. (We will continue to use the 1968 NYC population, but in fact the size of the population is essentially irrelevant.)
Complete and Enter the following commands.

```
> fractionOfPopulationWhichIsImmune := ??? :
> C := 2.00 :
> L := 0.13 :
> population := 7900000:
> r0 := population * fractionOfPopulationWhichIsImmune:
> Days := 135:
> EpidemicModeler(C, L, initialImmunes=r0, plots=infected, days=Days);
>
```

What did you find as the smallest fraction of the population that needs to be immune to measles to guarantee that no epidemic occurs? Place your answer below.
Answer:

Suppose we have a vaccine for the measles that is 95% effective. What percentage of the population would have to be inoculated in order to prevented an epidemic, i.e., to achieve "herd immunity"? How difficult do you think this would be to achieve? Place your answer below.
Answer:

• Rubella.

Repeat your estimation of the inoculation level necessary to achieve herd immunity for rubella ("German measles"), a disease for which b=0.66 and l=0.09 (approximate values). Continue to assume the existence of a vaccine which is 95% effective. Which disease seems easier to control, measles or rubella?
Enter your calculations and conclusions below.

```
> fractionOfPopulationWhichIsImmune := ???:
> C := 0.66 :
> L := 0.09 :
> population := 7900000:
> r0 := population * fractionOfPopulationWhichIsImmune:
> Days := 135:
> EpidemicModeler(C, L, initialImmunes=r0, plots=infected, days=Days);
>
```

Answer:

¤ Theoretical Extra Credit Question.

For each of the three diseases just considered you have estimated a number Smax such that if (S0/N)<Smax, then no epidemic occurs, while if (S0/N)>Smax, then an epidemic will occur (N is the total population). Do you see any relationship between your values for Smax and the corresponding values of C and L ? Such a relationship does exist, and can be obtained by analyzing the differential equation for I(t):

dI/dt = (C/N)*I(t)*S(t) - L*I(t).

What do you think the relationship is between Smax, C, and L? Verify your conjecture by analyzing the differential equation shown above. (Hint: If no epidemic occurs, then what can you say about the sign of dI/dt when t=0?)
Place your conjecture and verification below.
Answer:

LAB 10: A Discrete Price Model

□ NAMES:

□ PURPOSE
In this lab we consider a model for the evolution of prices in an economy with discrete trading times. Features of this model are investigated and compared with features of the continuous model of price evolution described in Calculus: Modeling and Application.

□ PREPARATION.
Study the material on the continuous and the discrete price models in Section 5.4 of Calculus: Modeling and Application. In particular, be sure you understand Checkpoints 3-6 and Exploration Activities 1 and 2. Read over these instructions before coming to the lab.

□ LABORATORY REPORT--General Instructions.
Note: Follow your instructor's instructions if they differ from those given here.

You are to write a self-contained, well-organized exposition and analysis of the issues raised in this laboratory. In particular, it should incorporate the answers to all the important questions raised during the lab session.

Two submissions will be made for this report: an initial draft, and a final revision. Comments made on the draft will help you prepare the revision, and only the revision will receive a grade. Moreover, the grade will be based solely on the written report--the instructor will look at other parts of the notebook only in unusual situations. You should therefore consider the entries you make in the Preliminaries and Project sections to be notes to help when later writing your report.

More specific instructions on the content of the report will be given at the end of the worksheet.

□ Maple Procedures. Please Enter the following line.

```
> with(lab10);
> # If an Error is an output of the above line, see your lab instructor.
```

Lab 10 Procedures define the one command that will be the central tool of this lab: PricePlotter. This produces various useful combinations of graphs for our price model. For more information use the command ?PricePlotter .

□ PRELIMINARIES.

¤ 1. The Discrete Price Model.

The following are the assumptions for our discrete price model, as described on pages 327 through 330 of Calculus: Modeling and Application:

The trading times occur at time intervals of length Deltat. Thus the trading times are t(k)=k*Deltat for k=0, 1, 2, ... The prices at these times will be denoted by p(k), k=0, 1, 2, ... (Note: In what follows we will always assume that Deltat equals one day.)

At the k-th trading time, demand is modeled by

D(k) = a + b*p(k), where a and b are constants, $b < 0$.

At the k-th trading time, supply is modeled by

Q(k) = c + d*p(k-1), where c and d are constants, $d > 0$.

The price p(k) must be set so that market clearing occurs. This means

D(k) = Q(k).

Explain (in terms of the nature of supply and demand) why b is negative and d is positive. Place your answers below.

Answer:

Using these assumptions it is shown on page 328 of Calculus: Modeling and Application that the price satisfies the following Recursion Relation: Activate the following command line.

```
> Result(1);
```

Moreover, this recursion relation can be shown to lead to a Closed Form Equation: Activate the following command line.

```
> Result(2);
>
```

The major steps in obtaining this closed form equation are discussed on pages 328 through 330 of Calculus: Modeling and Application.

Note. We use the recursion relation to generate graphs of p(k), but we use the closed form equation to explain the behavior of the graphs.

¤ 2. The Continuous Price Model.

The following are the assumptions for our continuous price model, as described on page 324 of Calculus: Modeling and Application:

Trades are being made at all times and buyers and sellers respond immediately to changes in the price. Price will be denoted as a function of time by p(t).

At time t, demand is modeled by
D = a + b*p(t), where a and b are constants, b < 0.

At time t, supply is modeled by
Q = c + d*p(t), where c and d are constants, d > 0.

The price p(t) must satisfy the following Initial Value Problem: Activate the following command line.

```
> Result(3);
```

Using these assumptions it can be shown that the price function is given by a Continuous Solution Function: Activate the following command line.

```
> Result(4);
>
```

The major steps in obtaining this continuous solution function are discussed on page 326 of Calculus: Modeling and Application.

Note: We use the continuous solution function to generate graphs of p(t).

You should compare the Continuous Solution Function to the Closed Form Equation obtained in the discrete case. What are the similarities and what are the differences? You should answer these and subsequent questions in the lab report that you write at the end of this Worksheet. However, you can "jot down" notes for later reference. Place your answers below.

Answers:

¤ 3. Maple Code for the Discrete Price Model.

We wish to study the discrete price model by changing the values of the parameters of the model and observing the effects on the graph of price verses time. We therefore need to generate price verses time graphs. We start by specifying values for the parameters in the model (a, b, c, and d), as well as specifying the number of days that we wish to observe the model. This is done in the next cell. Enter the following parameter specification cell.

```
> a :=12; b :=-1.5; c :=-1; d :=1; days :=20;
>
```

We now generate the prices on successive days, p(1), p(2), p(3), ... , by use of the recursion relation from Section 1.

• Maple Tip. Place this recursion relation inside a Maple proc using the format given below. This will automatically define a remember table. You then assign the initial value which is inserted into the remember table, and finally ask Maple to compute some values using the seq command.

• Warning!!! It is critically important to test a recursion relationship before running it too many times. For that reason you will first test your seq command with k running from 1 to 4.

```
> p := proc(k)
> p(k) := ??? *p(k-1)
> end:
> p(0) := 5:
> seq( p(k), k=1..4);
>
```

• Important!!! Your result should be a sequence of four numbers. If anything else appears, then do not execute the seq command for k larger than 4 until you fix the problem!!

If necessary, continue to correct and reEnter your seq (with k going up only to 4) until the resulting output is a list of four numbers. Correct and reEnter your "test" seq command above as needed.

If you do not have a problem... You are now ready to run your seq command for the full range of k values. Simply change k=1..4 to k=1..days, and then Enter the command. Enter your seq command with the full k range.

Now construct a data list pData with a seq command, and plot the result with a plot command. This is our desired graph. Enter the following seq and plot commands.

```
> pData := [seq([k, p(k)], k=0..days)]:
> plot(pData);
>
```

Although this series of commands is simple enough to run "as is," for your convenience we have constructed a new command PricePlotter which contains this code. The command allows us to include Maple "bells and whistles" for enhanced clarity and flexibility, and enables us to run the code with just one input command. Note, however, that the mathematical heart of this portion of PricePlotter consists of nothing more than the commands just discussed.

Enter the following command for a demonstration run of PricePlotter.

```
> a :=12: b :=-1.5: c :=-1: d :=1:
> PricePlotter(a, b, c, d);
```

Notice that the initial price has a default value of 5 and the number of days shown in the plot has a default value of 20. You can alter these values with the options InitialPrice=number and Days=number. The constant Gamma is given as 0.1700 . You can also control the portion of the graph that is plotted by the use of the PlotRange option. As an example, PlotRange=5..15 would plot the prices for days 5 to 15. Try out a plot using this option. Enter your command below.

```
> a :=12: b :=-1.5: c :=-1: d := 1:
> PricePlotter(a, b, c, d, PlotRange=5..15);
>
```

¤ 4. Maple Code for the Continuous Price Model.

We generate graphs for the continuous price model in much the same way as for the discrete model. We start by specifying values for the parameters in the model (a, b, c, d, and Gamma), as well as specifying the initial price p(0) and the number of days that we wish to observe the model. An example set of variables is given on the next line. Enter it.

```
> a :=12: b :=-1.5: c :=-1: d := 1: g :=0.17: p0 := 5: days := 20:
```

The desired plot will simply be a plot of the continuous solution function pc(t) from Section 2. (We use pc(t) rather than p(t) because we already have assigned p(...) to the discrete model.) We define pc(t) by first defining the term P (as it appears in Section 2). Complete, then Enter, the following commands.

```
P := ???; pc := t -> ???;
> P := (c-a)/(b-d): pc := t -> P+(p0-P)*E^(-g*(d-b)*t);
>
```

• Maple Tip. Maple does not know the symbol "e"--however, it does know "E" or exp(...).

We now graph pc(t) using a Plot command. This is the desired graph. Enter the following Plot command.

```
> plot(pc(t),t=0..days);
>
```

Although these commands are simple enough to run "as is," for your convenience we have included them in PricePlotter--they are accessed via the option Model=continuous. Although some Maple "bells and whistles" are included in the new command, note that the mathematical heart of this portion of PricePlotter consists of nothing more than the commands just discussed.

Enter the following command for a demonstration run of this portion of PricePlotter.

```
> a :=12: b :=-1.5: c :=-1: d := 1: gam :=0.17:
> PricePlotter(a, b, c, d, g=gam, Model=continuous);
```

As with the discrete model, the initial price has a default value of 5 and the number of days shown in the plot has a default value of 20. These quantities are controlled by the options InitialPrice and Days. You can also control the portion of the graph that is plotted by the use of the PlotRange option.

¤ 5. Maple Code for the Discrete and Continuous Models Together.

During the project we will want to place discrete price model and continuous price model graphs together in one plot. You know how to do this with a plots[display] command, but we have built this capability directly into PricePlotter; you simply include the option Model=both as shown in the following example. Enter the following commands.

```
> a :=12: b :=-1.5: c :=-1: d := 1: gam :=0.17:
> PricePlotter(a, b, c, d, g=gam, Model=both);
>
```

You can request more information about PricePlotter by executing the command ?PricePlotter. Type and Enter your command below.

□ PROJECT.

¤ 1. The Discrete Price Model--the First Look.

We wish to consider the trading environment where

a = 12, b = -1.5, c = -1, and d = 1.

Plot the price function for the first 20 days. What happens to the price as the number of trading sessions increase? How could you have predicted the equilibrium price in advance? You should answer these and subsequent questions in the lab report that you write at the end of this worksheet. However, you can "jot down" notes here for later reference. Enter your commands and place your discussion below.

```
> a :=???: b :=???: c :=???: d := ???:
> PricePlotter(a, b, c, d);
>
```

Discussion:

¤ 2. The Discrete Price Model--Varying the value of d.

Now experiment with the plot by increasing d slowly through the range from 1 to 2. What does an increase in the parameter d say about the attitudes of the suppliers? Describe what happens and give an intuitive explanation for what you have observed. Enter your commands and place your discussion below.

```
> a :=12: b :=-1.5: c :=-1: d := ??? :
> PricePlotter(a, b, c, d);
>
```

Discussion:

Using the closed form formula for the price p(k), explain why the behavior observed as d increases is totally expected. Place your discussion below.

Discussion:

¤ 3. Comparison with the Continuous Price Model.

Experiment with the combined discrete and continuous plots, determining the behavior of the plots as d increases from 1 to 2. Set the value of g in the continuous model to be 0.17 Explain the differences between the two models. How could all the changes in behavior have been predicted from the closed form equation for the discrete model and the solution formula for the continuous model? Enter your commands and place your answers below.

```
> a :=12: b :=-1.5: c :=-1: d := ???: gam :=0.17:
> PricePlotter(a, b, c, d, g=gam, Model=both);
>
```

Discussion:

□ LABORATORY REPORT--Specific Instructions.

Write a self-contained exposition of the issues raised in the lab, along with your observations, explanations, and conclusions. Support your work as necessary with Maple generated graphs. Answers to the specific questions should be embedded in your exposition but should not substitute for it. As usual, the clarity and quality of your explanations is as important as the content of the report.

Place your report below:

□

LAB 11: Trigonometric Functions

□ NAMES:

□ PURPOSE.
The purpose of this lab is to gain experience with the sine and cosine functions--the fundamental periodic functions--and to discover an identity that is important in science and engineering.

□ PREPARATION.
Read the first nine pages of Chapter 6 in Calculus: Modeling and Application, and work Checkpoints 2, 3, 5, and Exploration activity 3. Read these instructions before coming to the lab, and bring them with you.

□ LABORATORY REPORT INSTRUCTIONS.
For this project you are not required to submit a formal report. Rather you are asked to submit a copy of this Worksheet with the questions answered in the indicated locations. Only one submission will be made. Write in complete sentences and be sure the sentences connect together in coherent ways.

□ Maple Procedures. Please Enter the following line.

```
> with(lab11);
> # If an Error is an output of the above line, see your lab instructor.
```

The Maple procedures for Lab 11 define two commands that will be the central tools of this lab: TwoPlot and TwoPlotD. TwoPlot draws two functions in one graph, automatically assigning different colors to the curves. TwoPlotD does the same, except that it also includes the graphs of the derivative curves.

□ PROJECT.

¤ 1. The basic sine and cosine functions.

We start our investigation of trigonometric functions with the sine and cosine. Enter the following function definition commands.

```
> f := t -> cos(t); g := t -> sin(t);
```

It would be easy enough for you to instruct Maple to graph these two functions in one plot. However, we have put together a simple routine that graphs two functions and their derivatives in one plot, and also varies the styles of each graph so that they are readily distinguishable from each other. This new routine is called TwoPlotD. Enter ?TwoPlotD to find out more about this command. Then use it with f and g over the interval 0 <= t <= 4*Pi. Enter your commands below.

```
> TwoPlotD(f,g,0..4*3.14);
>
```

In the plot just generated, some of the curves should overlay each other (if this is not clear, then enlarge your plot)--this means that the functions corresponding to these curves are the same! Moreover, other curves appear to be mirror images of each other across the horizontal axis--this means that the functions corresponding to these curves are the negatives of each other! Write down the relationships that you see among the curves, and then give the corresponding function identities. Explain these identities in terms of your reading in Chapter 6. Place your answers below.

¤ 2. Varying the period of the sine and cosine.

We now generalize slightly the functions f(t) and g(t) from the previous section: we add in a multiplicative factor k on the independent variable t. Enter the following function definition commands.

```
> f := t -> cos(k*t);  g := t -> sin(k*t);
>
```

Execute TwoPlotD for f and g, varying the constant k. How does the value of k affect the periods of the functions and derivatives? (The period of a function is the length on the t-axis for a full repetition of the function to occur) As done in Section 1, write down the relationships that you see among the curves, and then give the corresponding function identities. Explain these identities in terms of the Chain Rule. Place your plots and answers below.

```
> k := ???;
> ???
>
```

¤ 3. Varying the amplitude of the sine and cosine.

We now generalize slightly the functions f(t) and g(t) from the Section 1 in a different direction: we add in a multiplicative factor a on the dependent variable. Enter the following function definition commands.

```
> f := t -> a*cos(t);  g := t -> a*sin(t);
>
```

Execute TwoPlotD for f and g, varying the constant a. How does the value of a affect the amplitudes of the functions and derivatives? (The amplitude of a periodic function is its maximum value.) As done in Section 1, write down the relationships that you see among the curves, and then give the corresponding function identities. Discuss briefly what differentiation rules are illustrated by your identities. Place your answers below.

```
> a := ???;
> ???
```

¤ 4. Displacing the sine and cosine.

We investigate yet a third generalization of the functions f(t) and g(t) from the Section 1: we add a constant d to the independent variable. Enter the following function definition commands.

```
> f := t -> cos(t + d);  g := t -> sin(t + d);
>
```

Execute TwoPlotD for f and g, varying the constant d. The effect of d on the graphs may be difficult to see; for that reason, we have included an "animation" in the next region of commands. The effect of the d will then be very clear. Our command will generate plots for six values of d, increasing in increments of 0.25.
Enter the following command, then run the animation.

```
> for k from 1 to 6 do
>   d := 0.25 * (k-1):
>   frame[k] := TwoPlotD(f,g,0..4*3.14):
>   printf(`.`);      # This prints out one dot per iteration.
> od:
> plots[display]([seq(frame[k], k=1..6)], insequence=true);
>
```

Explain how the graph of the function changes as d varies. As done in Section 1, write down the relationships that you see among the curves, and then give the corresponding function identities. What differentiation rule is illustrated by your identities? Place your answers below.

¤ 5. Putting it all together.

We can, of course, vary the period, vary the amplitude, and vary the displacement all at once. Enter the following function definition commands.

```
> f := t -> a*cos(k*t + d);
> g := t -> a*sin(k*t + d);
>
```

Execute TwoPlotD once or twice for f and g, varying the constants a, k, and d. Are you getting any surprising effects? Place your answers below.

```
> a := ???;  k := ???;  d := ???;
> TwoPlotD(???);
>
```

We now wish to focus just on the function g(t) = a*sin(k*t + d). For that reason we turn to the new command TwoPlot (no final D) in the following form. Enter the following command.

```
> TwoPlot(g, 0..4*3.14);
```

We will determine some surprising alternate expressions for this curve in the next two sections.

¤ 6. Sums of sines and cosines.

We now turn to another collection of periodic functions: sums of sines and cosines. Enter the following function definition command.

```
> G := t -> A*cos(k*t) + B*sin(k*t);
>
```

Plot G(t) using TwoPlot for a few values of A, B, and k. How do the shapes of your plots compare to the plots of g(t) = a*sin(k*t + d) that you obtained in the previous section? Enter your commands and answers below.

```
> ??? ??? ???
> ???
>
```

¤ 7. A comparison.

For convenience, here are the functions from the previous two sections that we are interested in.

g := t -> a*sin(k*t + d)
G := t -> A*cos(k*t) + B*sin(k*t)

We wish to more formally compare g(t) with G(t). The following region allows you to vary the five parameters a, k, d, A, and B, and then graph the two functions g(t) and G(t) in the same plot. Enter the following set of commands.

```
> A :=3; B :=4; k :=1;
> a :=2; d :=1.5;
> TwoPlot(g, G, 0..4*3.14);
>
```

Use this region to find values of a and d so that g(t) and G(t) seem to agree, and record the values in the cell below. Continuing to keep k=1, repeat this procedure with two other pairs of values for A and B, recording your results below. (Hints: Start with a=1 and d=0. Then by how much do you need to multiply the blue curve to get a height equal to the red curve? By how much do you need to shift the blue curve to the left in order to match the red curve?) Enter your commands and results below.

• Maple Tip. You can obtain the coordinates of any point by clicking anywhere inside the plot, the coordinates of the location of the cursor will show up in the lower left corner of the screen.

Results:

If k=1, A= 3, B= 4, then a= ___, and d= ___.
If k=1, A= ___, B= ___, then a= ___, and d= ___.
If k=1, A= ___, B= ___, then a= ___, and d= ___.

¤ 8. An analysis.

Obtain descriptions of a and d in terms of A and B. For a, look for a formula in terms of A and B (you might be able to do this directly from your data). Try to identify d as an angle in a right triangle. After you obtain your descriptions, check them against your experimental data in Section 7. (Hint: Apply the trigonometric identity

sin(s+t) = sin(s)*cos(t) + cos(s)*sin(t)

to the function g(t).) Place your answers below.

Suppose the value of k is something other than 1. Does this change your descriptions or formulas for a and d? Experiment (i.e., using TwoPlot) to check that your answer is correct. Place your answers below.

State clearly the conditions under which

g(t)= a*sin(k*t + d) and G(t) = A*cos(k*t) + B*sin(k*t)

are the same function. Place your answers below.

LAB 12A: Pendulum Motion

□ NAMES:

□ PURPOSE.
In this lab we use numerical solutions of the pendulum equation for various pendulum lengths to determine the relationship between length and period. There are two other ways of describing the relationship between period and length: direct observations of periods (i.e., experimentation as might be done in a physics lab) and the exact solution of an approximate pendulum equation (as described in Calculus: Modeling and Application). Our ultimate purpose is to compare these three methods for solving the same problem.

□ PREPARATION.
Study pages 371-373, and 377-383 in Chapter 6 of Calculus: Modeling and Application, work Exploration Activity 3 on page 371, and Checkpoints 1-6 in section 6.3.

□ LABORATORY REPORT INSTRUCTIONS.
For this project you are not required to submit a formal report. Rather you are asked to submit a copy of this Worksheet with the questions answered in the indicated locations. Only one submission will be made. Write in complete sentences and be sure the sentences connect together in coherent ways.

□ Maple Procedures.
You need to Enter the next line to run this lab.

```
> with(lab12a);
> # If an Error is an output of the above line, see your lab instructor.
```

Lab 12A procedures define three commands that will be central tools of this lab: pendulum, pendulumPlot, and periodFinder. Pendulum generates numerical solutions to the pendulum equation, pendulumPlot plots the data generated by pendulum, and periodFinder prints out portions of the pendulum data to allow a determination of the period. These commands will be further explained when needed.

□ PRELIMINARIES.

¤ 1. The Basic Setup.

In Calculus: Modeling and Application we developed the two differential equations and the two initial conditions used to model the motion of a pendulum (Exploration Activity 3). For convenience, we repeat them here:

dTheta/dt = v,   Theta(0) = Pi/6
dv/dt = -(g/L)*sin(Theta),   v(0)=0

Recall that

Theta(t) is the angle of the pendulum at time t,
v(t) is the velocity of the pendulum at time t,
i.e., the derivative of Theta(t),
L is the length of the pendulum, and
g is the acceleration due to gravity.

In this lab we will approximate solutions to this "double" initial value problem by using Euler's Method simultaneously on the two initial value problems for Theta and v--we must work on both equations simultaneously because the variables Theta and v are intertwined.

Starting with a pendulum length L=len=100 cm, and a time step of Deltat=.005 sec, we will apply the Euler Method n=600 times; the values of the quantities Theta and v so produced at time t=k*dt will be denoted by Theta(k) and v(k). We will repeat this procedure for various pendulum lengths L and then analyze the results to determine a relationship between pendulum length and period.

¤ 2. Deriving the "Updating Equations" for Euler's Method.

Given the values Theta(k) and v(k), we wish to derive equations for the next set of values Theta(k+1) and v(k+1). To do this we use Euler's Method, the central idea being the replacement of derivatives with difference quotients in the two differential equations from the previous section. At the k-th step define

DeltaTheta = Theta(k+1)-Theta(k) and Deltav = v(k+1)-v(k).

Using Deltat as the time step, replace the derivatives with difference quotients (DeltaTheta /Deltat and Deltav/Deltat) in the two differential equations from Section 1. Compute your results on paper, then check them against the answers revealed by activating the Result(1) line.

• Learning Tip. Resist activating the line until you at least attempt to write the answer. Writing things out by hand is an important technique for learning.

```
> Result(1);
```

Solve these two equations for Theta(k+1) and v(k+1). Compute your results on paper, then check them by activating the next line.

```
> Result(2);
>
```

(You have just solved Exploration Activity 3.)

We have the starting values Theta(0) and v(0) --these are just the given initial conditions. Now, however, we can use the equations just derived to go to the first set of computed values, Theta(1) and v(1) , and from these to the second set of computed values, Theta(2) and v(2), and so on. This is how Euler's Method will be implemented in the next section.

¤ 3. The Maple Commands--Implementing Euler's Method.

For various values of L=len, the length of the pendulum, we wish to estimate the period. To do this it was convenient to construct the following Maple commands: pendulum, pendulumPlot, and periodFinder. Find out about these new commands by the use of ?name. Enter your commands below.

As an example of the first two of these commands, the statements below will generate data for a pendulum of length len=100 cm, using n=600 steps with a step size of Deltat=.005 sec. The data is then used to generate two plots: angle verses time, and velocity verses time. Enter the following region and examine the output.

```
> len := 100.:   n :=600:   Deltat :=.005:
> pendulumData :=pendulum(len, n, Deltat):
> pendulumPlot(pendulumData);
>
```

□ PROJECT.

¤ 1. Determining the period for various lengths of a pendulum.

We want to estimate the period of the pendulum for various values of the length L=len. One way to do this is to estimate the length of half of a full cycle. Examine the graph of the angle Theta that was obtained at the end of the Preliminaries section: the curve is decreasing during the first half of the cycle, and then increasing during the second half. This means that at the half way point in a cycle, the slope of the Theta curve--i.e., the velocity v--will turn from ________ to ________ (Fill in the blanks). Determining where the velocity curve makes this change is how we will compute the (half) period for various values of the pendulum length.

Start with the length L=len=100, and plot the graphs of Theta and v by executing the following region.

• Maple Tip. This is the same command that you used in the Preliminaries section. From the graph obtained in that section you should see how to reduce the value of n from 600 to something considerably lower and still reach the desired transition point for the velocity curve. Make such a reduction--it will significantly speed up your computations.

Execute the following region (first modifying the n value if so desired):

```
> len :=100:   n :=600:   Deltat :=.005:
> pendulumData :=pendulum(len, n, Deltat):
> pendulumPlot(pendulumData);
```

From the graphs you obtain, estimate the time t at which half an oscillation is completed (the half period). Remember, this occurs when the variable v turns from negative to positive. Call the t-value you obtain tEst (for t-estimate) and place it into the following region. Executing this cell will print all of the v values close to the t value that you choose. Now read off from the list the best t value for the (full) period. Write your t value down on a piece of paper and save it for later use. Remember that this is the value of the period that corresponds to the length len=100 cm. Complete and Enter the following commands.

```
> tEst := ???;
> periodFinder(pendulumData, tEst);
>
```

Now repeat these procedures for at least 10 values of len between 0 and 100, including at least three values of len less than or equal to 10. As len decreases, you will be able to decrease the value of n (why?). However, in order to get satisfactory results, in some instances you may also have to decrease the step size Deltat (which will then require a corresponding increase in n). Place your computations below.

```
> len := ???:  n := ???:  Deltat := ???:
> pendulumData := pendulum(len, n, Deltat):
> pendulumPlot(pendulumData);
> tEst :=  ???:
> periodFinder(pendulumData, tEst);
>
```

Record all of your results, i.e., ordered pairs [length, period], in the following data list. Enter the cell when your computations are finished. Complete and Enter the following data list.

```
> periodData := [[100,???], [???,???], ??? ];
>
```

¤ 2. Period as a function of pendulum length--fitting the graph.

Use the appropriate Maple commands to obtain a graph of the periodData list; call this graph periodPlot. Enter your Maple commands below.

```
> periodPlot := ???:";
>
```

Given the material you've read in Calculus: Modeling and Application, did you obtain the type of curve that you expected? What sort of function of the length L do you think would produce such a graph? Here are some possible choices:

A*L^2 , A*sin(w*L) , A* sqrt(L), A*e^(k*L) , A*log(k*L)

Place your answer below as a Maple function definition statement for the function period=period(L). Complete and Enter the following statement.

```
> period := L -> ???;
```

We wish to vary the value of the parameter A that appears in the definition of the function period until we obtain the best fit to the data in periodData. We can do this by placing the graph of period over the graph of periodData. We then adjust the value of A until we get good agreement of the two graphs. The following Maple code produces the desired plots; you merely supply the value of A. Execute the following region with various values of A until a good fit is achieved:

```
> A := ???;
> plots[display]([periodPlot, plot(period(L),L=0..100, color=COLOR(RGB,1,.3,0))]);
>
```

When fitting a function to a data set, it is reasonable to check the fit by looking at the differences between the actual data and the modeling function. The collection of such differences is known as the residual. A "good" fit will result in a "random" looking residual; a "bad" fit will result in some "non-random" pattern emerging in the residual. The following region computes and plots the residuals of our Euler Method data as modeled by the function period(L). Enter the following two regions.

```
> N := nops(periodData);  # Number of elements (operators) in periodData.
> residuals :=evalf([seq([periodData[k][1],
>                          periodData[k][2] - period(periodData[k][1])], k=1..N)]);
> plot(residuals, style=POINT, symbol=CIRCLE, color=COLOR(RGB,1,0,0));
>
```

Comment on the graph of the residual. Do you see a pattern? If so, to what might you attribute this behavior? Use the residual graph to make fine tunings in your value of A. Place your answers below.

¤ 3. Comparison with the Linearized Model.

On page 382 in Chapter 6 of Calculus: Modeling and Application, pendulum motion is modeled by the use of an approximate pendulum equation, i.e., we replace

d^2Theta/dt^2 = -(g/L)*sin(Theta)

by the "linearized" equation

d^2Theta/dt^2 = -(g/L)*Theta

This equation is pleasant to deal with because it can be solved in terms of our well-known sine and cosine functions--all solutions for this equation are of the form

Theta(t) = A*sin(sqrt(g/L)*t) + B*cos(sqrt(g/L)*t)

From this formula it is easy to determine the period of Theta = Theta(t) as a function of L. Using the value g=980.7 cm/sec^2, write your formula in the way that most nearly matches period(L), the formula determined via the numerical solutions in the previous section. Determine the period, then check your answer by activating the following line.
Answer:

```
> Result(3);
>
```

How well does the period formula obtained via the linearized model compare with the period formula obtained via the numerical solutions of the original IVP? Place your answer below.

¤ 4. Comparison with experimental data.

In this section we compare the period formula obtained from the numerical solutions with periods measured experimentally. During a class session you might have taken such measurements yourself; if so, then use your data in this part of the lab. If not, then use the data provided in the next cell. The data provided consists of 18 pairs, each pair consisting of a pendulum length and the corresponding cycles per minute. This data--collected by a Project CALC instructor--was obtained in a rather crude fashion: a fat plumb bob was tied to a string and suspended from a nail in a door frame, and the number of cycles per minute was counted for each of 18 lengths of the pendulum. The length of each pendulum was determined by a measurement to the (roughly estimated) balance point of the plumb bob. Enter the following (or your own) data list.

```
> cyclesData1 :=[[190, 21.75], [180, 22.5 ], [170, 23.0 ], [160, 23.75], [150, 24.5],
> [140, 25.4 ], [130, 26.25], [120, 27.25], [110, 28.5 ], [100, 30. ], [ 90, 31.5 ], [ 80, 33.2 ], [ 70, 35
> .5 ], [ 60, 38.3 ], [ 50, 41.6], [ 40, 46.5 ], [ 30, 54. ], [ 20, 66. ]];
>
```

If the data list just entered consists of cycles rather than periods (as is the case with the data we have provided), then it must be converted to periods. To do this, recall that the period is merely the time per cycle, which is the inverse of cycles per time. The following command converts cycles to periods in just this fashion. Enter the following cell.

Note: If you are using your own data, and your data is already in the form [length, period], then you should just label your data as periods0 and skip the following command.

```
> periodData1 := [seq([cyclesData1[k][1], 60/cyclesData1[k][2]],
>                   k=1..nops(cyclesData1))];
>
```

Use the appropriate Maple commands to obtain a graph of the periods0 list; call this graph periodPlot0. Enter your Maple commands below.

```
> periodPlot1 := ???:";
```

The plot obtained above should look similar to the plot obtained earlier via numerical methods. As done then, we define a function period0(L) which we intend to fit to the curve. Use the same function as done for the numerical solutions, except with a B in place of the A. Complete and Enter the following function definition statement.

```
> period1 := L -> ???;
>
```

We wish to vary the value of the parameter B that appears in the definition of the function period1 until we obtain the best fit to the data in periodData1. We can do this by placing the graph of period1 over the graph of periodData1. We then adjust the value of B until we get good agreement of the two graphs. The following Maple code produces the desired plots; you merely supply the value of B. Execute the following commands with various values of B until the desired fit is achieved:

```
> B := ???:
> plots[display]([periodPlot1,plot(period1(L),L=0..190,color=COLOR(RGB,1,.3,0))]);
>
```

How well does the resulting formula for period1(L) match the formula for period(L)? Enter your answer below.
Answer:

We again check our fit by looking at the residual, the collection of differences between the actual data and the modeling function. A "good" fit will result in a "random" looking residual; a "bad" fit will result in some "non-random" pattern emerging in the residual. The following cell computes and plots the residuals of our experimental data as modeled by the function period1(L). Enter the following two regions.

```
> N := nops(periodData1):
> residuals1:=evalf([seq([periodData1[k][1],
>                 periodData1[k][2] - period1(periodData1[k][1])], k=1..N)]);
> plot(residuals1, style=POINT, symbol=CIRCLE, color=COLOR(RGB,1,0,0));
>
```

Comment on the graph of the residual. Do you see a pattern? If so, to what might you attribute this behavior? Use the residual graph to make fine tunings in your value of B. Place your answers below.
Answer:

¤ 5. Analysis of the various methods.

Three different models have been developed to relate the period of a pendulum to its length: the numerical model derived from Euler's Method computations, the analytic model obtained from solving the linearized pendulum equation, and the experimental model derived from direct measurements of pendulum motion. In this section we focus on what each approach contributes to our understanding of pendulum motion.

(a) Which approach provides the firmest grounding in "physical reality," i.e., if you wished to convince someone that our square root model accurately portrays the period of a pendulum, which model would you refer to? Place your answer below.

(b) Which approach provides the best "theoretical justification" for using a square root model to portray the period of a pendulum, i.e., if you wished to convince someone that the use of a square root model is not just an empirical choice, but is truly intrinsic to the problem, which model would you refer to? Place your answer below.

(c) Which approach most clearly "tests the basic model" we have derived for the period of a pendulum, i.e., if you wished to convince someone of the reasonableness of the underlying assumptions we have made and the IVPs to which they led, which model would you refer to? Place your answer below.

(d) Is it valuable to work through all three approaches to modeling the period of a pendulum, or is one (or more) of the models simply redundant, i.e., is the extra (incremental) understanding achieved from each additional approach worth the time and effort expended on it? To bring the question closer to home, do you feel that your understanding (and belief in) the square root model was significantly advanced by each of the solution methods used in this lab? Place your answer below.

□

LAB 12B: Ideal Spring Motion

□ NAMES:

□ PURPOSE
As described in Calculus: Modeling and Application, the oscillation of an ideal spring/mass system depends only on the ratio of the mass and the spring constant, along with the starting displacement. In this lab we focus on determining the relationship between these quantities and the period and amplitude of the solution curves. We will assume that there is no significant retarding frictional force, i.e., that we have an undamped spring.

The dependence of the period and amplitude on the starting quantities will be studied from three different viewpoints: numerical, algebraic (symbolic), and empirical.

□ PREPARATION.
Prior to the laboratory session you should study Sections 6.1 and 6.2 of Calculus: Modeling and Application.

Read the Preliminaries, writing out answers to the questions which are asked. During the lab session you will then be able to check your work against those obtained by activating command lines in the worksheet and move quickly to the Project.

□ LABORATORY REPORT INSTRUCTIONS.
For this laboratory you are not required to submit a formal report. Rather you are asked to submit the following:
¤ An electronic copy of this Worksheet with the questions answered in the indicated locations. Only one submission will be made. Write in complete sentences and be sure the sentences connect together in coherent ways.
¤ A filled-in paper copy of the Laboratory Summary Sheet. This is the last section of the laboratory worksheet.

□ Maple Procedures.
Be sure to initialize the Worksheet before beginning the lab.

```
> with(lab12b);
> # If an Error is an output of the above line, see your lab instructor.
```

□ PRELIMINARIES.

¤ 1. A Differential Equation Model for an Ideal Spring/Mass System.

In this lab we consider the movement of a mass on a spring, as described in Chapter 6 of Calculus: Modeling and Application. To reinforce the physical origin of the model, the following six steps lead you through the derivation of the initial value problem which models such a system. This parallels the discussion in the subsections "The Spring-Mass System" and "The Initial Value Problem" of Sections 6.2 of Calculus: Modeling and Application.

What follows is a rendering of Figures 6.9 and 6.10, page 367 of Calculus: Modeling and Application.

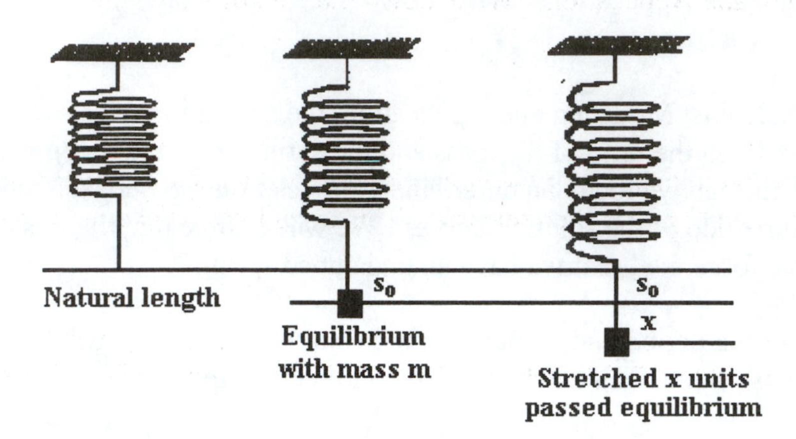

• Step 1. Suppose a mass m stretches a spring by s0 units from its natural length. Then, according to Hooke's Law, the force exerted by the spring on the mass is -k*s0, where k is the spring constant (we will denote downward forces as positive, and upward forces as negative). However, the force of gravity, m*g, will balance the spring force, so that 0 = ..., which gives .....=......Write answers to the questions on paper, then activate the line which follows.

```
> Ans(1);
>
```

• Step 2. Take the spring mass system from Step 1, stretch it beyond the equilibrium s0 that it had achieved in Step 1, and then let go. At any time t let x(t) be the distance beyond the equilibrium s0 to which the mass is pulled (downward displacement is positive). Discounting retarding frictional forces, there will be only two significant forces acting on the mass. Write answers to the questions on paper, then activate the line which follows.

1. The force of gravity = ..., which from Step 1 can be alternately expressed as ......
2. The force of the spring = ...

```
> Ans(2);
```

• Step 3. Thus, at time t, the total force on the mass will be F = ... Check your answer by activating the following line.

```
> Ans(3);
```

• Step 4. By Newton's Second Law of Motion, a force F on a mass m will cause an acceleration a, where the three quantities are related by the formula F = ... Moreover, the acceleration can be expressed in terms of the position x=x(t) of the mass as a = ... Check your answers by activating the following line.

```
> Ans(4);
```

• Step 5. Combining the results of Steps 3 and 4 yields a differential equation model for the motion of a spring with damping. How does this compare with the last equation on page 357 of Calculus: Modeling and Application? Write down the equation on paper, then activate the following line.

```
> Ans(5);
```

• Step 6. Assume that initially (t=0) the spring is stretched a distance x0 beyond the equilibrium it had achieved in Step 1, and is then released. This gives two initial conditions that must be satisfied by x(t):

x(0) = ..., and
v(0) = dx/dt(0) = ...,

where v=v(t) is the velocity of the mass at time t. Check your answers by activating the following line.

```
> Ans(6);
> # Scroll down with the down arrow.
```

The differential equation of Step 5, combined with the initial conditions of Step 6, gives a second-order initial value problem (IVP) model for the motion of an ideal spring-mass system.

¤ 2. Euler's Method for the Ideal Spring/Mass Model.

Recalling the methods employed in the subsection "A Numerical Solution: Euler's Method of Section 6.2 of Calculus: Modeling and Application, we convert our second-order IVP model into a system of two coupled first-order IVPs. To do this we introduce the velocity function v(t) = dx/dt. Then x and v must satisfy the following system of IVPs:

dx/dt = v, x(0) = x0
dv/dt = - (k/m)*x, v(0) = 0

Do you see how these equations follow from the IVP given in Steps 5 and 6 of the previous section?

Recall that x(t) is the (downward) displacement from the system's equilibrium position at time t,
v(t) is the velocity of the mass at time t, i.e., the derivative of x(t),
x0 is the initial displacement of the mass,
m is the mass at the end of the spring,
k is the spring constant from Hooke's Law,

We will begin the Project by approximating solutions to this "double" initial value problem by using Euler's Method simultaneously on the two initial value problems for x and v--we must work on both equations simultaneously because the variables x and v are intertwined. (This is the vector Euler's Method, first considered in the lab on epidemics.)

Starting with specific values for x0, m, k, and Deltat (the time step size), we will apply the Euler Method between 500 and 1000 times; the values of the quantities x and v so produced at time t=k*Deltat will be denoted by x(k) and v(k). We will repeat this procedure for various values of x0, m, and k in order to determine the effects on the amplitude and period of the spring oscillation which are predicted by our spring/mass model.

¤ 3. Deriving the "Updating Equations" for Euler's Method.

Given the values x(k) and v(k), we wish to derive equations for the next set of values x(k+1) and v(k+1). To do this we use Euler's Method, the central idea being the replacement of derivatives with difference quotients in the two differential equations from the previous section. At the k-th step define

Deltax = x(k+1)-x(k) and
Deltav = v(k+1)-v(k).

Replace the derivatives with difference quotients (Deltax /Deltat and Deltav/Deltat) in the two differential equations from Section 1. Write your results on paper, then check them by activating the following line.

```
> Ans(7);
```

Solve these two equations for x(k+1) and v(k+1). Write your results on paper, then check them by activating the following line.

```
> Ans(8);
>
```

We have the starting values x(0) and v(0)--these are simply the initial conditions. Now, however, we can use the equations just derived to obtain the first set of computed values, x(1) and v(1), and from these to obtain the second set of computed values, x(2) and v(2), and so on. This is how Euler's Method will be implemented in the next section.

¤ 4. The Maple Commands--Implementing Euler's Method.

For various values of the constants k, m, and x0 we wish to use Euler's Method to obtain numerical solutions for the spring/mass model. To do this conveniently, we have constructed the Maple command IdealSpring. Find out about this new command by the use of ?name. Enter your commands below.

As an example of the command, the statements below will generate a plot of displacement verses time for a spring/mass system with k=4, m=.1, and x0=.15, using n=600 steps with a step size of Deltat=.003. Activate the next command line and examine the output.

```
> k :=4:  m :=.1:  x0 :=.15:  n :=600:  Deltat :=.003:
> idealSpring(k,m,x0, n, Deltat);
```

The graph of the displacement x=x(t) generated above looks very much like the graph of a well-known function. What is the function we are referring to? Place your guess below.

□ PROJECT.

¤ 1. Numerical solutions with varying initial displacements x0.

From the development in Section 1 of the Preliminaries, the IVP model for our system is:
d^2x/dt^2 + (k/m)*x = 0, where x(0) = x0, and dx/dt(0) = 0.

In this subsection we will study this model numerically by using Euler's Method to generate numerical solutions for a variety of values of k/m and x0. We will take special notice of the amplitude and period of the solution curves.

Let's begin with the example used in Section 4 of the Preliminaries. We repeat below the relevant Maple commands from that section. Enter the following commands.

```
> k :=4:   m :=.1:   x0 :=.1:   n :=600:  Deltat :=.003:
> idealSpring(k,m,x0, n, Deltat);
```

Determine the amplitude and the period of this solution curve, and record your answers as the first row in the table below, as well as in your (paper copy of the) Laboratory Summary Sheet. Place your answers below.

• Maple Tip. To determine the amplitude and the period it will be useful to place the tip of the cursor near the relevant point of the graph and click the left button of the mouse. The coordinates of the location of the cursor will appear in the lower left corner of the screen.

x0	amplitude	period
???	???	???
???	???	???
???	???	???

Now change the value of the initial displacement x0, rerun the idealSpring commands with this new value, and record the amplitude and period into the table above and into your Laboratory Summary Sheet. Repeat this for different values of x0 until you see the consequence (or non-consequence) of changes in x0 on the amplitude and the period of the spring oscillation. Record your answers into the table and the Laboratory Summary Sheet.

In what way does the value of x0 affect the amplitude of the spring oscillation? Is your answer surprising, and does it match your intuitive understanding of the motion of a spring? Place your answer below and in the Laboratory Summary Sheet.

In what way does the value of x0 affect the period of the spring oscillation? Does your answer match your intuitive understanding of the motion of a spring? Place your answer below and in the Laboratory Summary Sheet.

¤ 2. Numerical solutions with varying ratios k/m.

We continue with an ideal spring/mass system, i.e., with no damping force of air resistance. The IVP model for our system is:

d^2x/dt^2 + (k/m)*x = 0,   where x(0) = x0, and dx/dt(0) = 0.

Notice from this equation that the quantities k and m only appear once, in a ratio k/m. For this reason, we will study the dependence of the amplitude and the period on the ratio k/m, not on the individual quantities k and m.

Let's return again to the example used in Section 4 of the Preliminaries. We repeat below the relevant Maple commands from that section. Enter the following commands.

```
> k :=4.:  m :=.1:  x0 :=.15:  n :=600:  Deltat :=.003:
> idealSpring(k,m,x0, n, Deltat);
> k/m;
>
```

Determine the amplitude and the period of this solution curve, and record your answers as the first row in the table below, as well as in your Laboratory Summary Sheet. Place your answers below.

k/m	amplitude	period
???	???	???
???	???	???
???	???	???

Now change the value of the ratio k/m, rerun the idealSpring commands with this new value, and record the amplitude and period into the table above and into your Laboratory Summary Sheet. Repeat this for different values of k/m until you see the consequence (or non-consequence) of changes in k/m on the amplitude and the period of the spring oscillation. Record your answers into the table and the Laboratory Summary Sheet.

• Note: If you intend to complete the optional Section 3 below, then you should record amplitudes and periods for at least five or six different values of k/m.

In what ways do the values of k and m affect the amplitude of the spring oscillation? Do your answers match your intuitive understanding of spring motion? (Note that a large value of k would correspond to a spring with a "strong" pull-back force. This is a consequence of Hooke's Law, F=-k*x.) Place your answer below and in the Laboratory Summary Sheet.

In what ways do the values of k and m affect the period of the spring oscillation? Do your answers match your intuitive understanding of spring motion? Place your answer below and in the Laboratory Summary Sheet.

¤ 3. Period as a function of k/m. (Optional)

We can do better than the qualitative observations that you probably gave in answer to the last question in the previous subsection: we can numerically determine an equation which expresses the period of the spring motion as a function of the ratio k/m. From such an equation we will know precisely how the stiffness of the spring (as measured by the spring constant) determines the period of the oscillation.

We first construct a data set periodData consisting of ordered pairs [k/m, period] taken from the table you completed in the previous subsection. Complete and Enter the following data list of pairs [k/m, period].

```
> periodData := [ [.., ..], ..];
```

• Maple Tip. To gain a feel for the data, plot it with a plot command. (The options style=POINT, symbol=CIRCLE, and color=red will make the points easier to see).
Enter your command below.

```
> plot(periodData, style=POINT, symbol=CIRCLE, color=red);
>
```

Do you think this might be exponential decay? Well, you can test your conjecture with a semi-log plot. If the data is decaying exponentially, what should a semi-log plot produce for you? Answer: ???. To produce a semi-log plot, first use the command semiLogList to generate the data. This command starts with a list of the form [[x1,y1], [x2,y2], ...] and acts as follows.

semiLogList([[x1,y1], [x2,y2], ...]) = [[x1,log(y1)], [x2,log(y2)], ...].

The log of each of the second coordinates has been taken. The semiLogList command has been supplied for this lab. Enter the following line.

```
> semiLogData := semiLogList (periodData);
```

Now, use the plot command to plot semiLogData. It is recommended that you use the plot options style=point, symbol=circle, color=red, title=`Semi-Log Plot of period data`, and labels=[`k/m`,`ln(per.)`] to add clarity to the plot.) Enter your plotting command below.

```
> plot(semiLogData, style=point, symbol=circle, color=red,
>       title=`Semi-Log Plot of periodData`, labels=[`k/m`,`ln(per.)`]);
>
```

On the basis of your semi-log plot, do you think the relationship is exponential? Place your answer below.

Now let's test for a power relationship between k/m and period. Figure out why the use of a log-log plot can test for a power relationship. What type of test can we use for this? First, construct the data with the logLogList command supplied with this lab.

logLogList([[x1,y1], [x2,y2], ...]) = [[log(x1), log(y1)], [log(x2), log(y2)], ...]

This takes logs of both coordinates. Enter the following line.

```
> logLogData := logLogList(periodData);
> plot(logLogData, style=point,symbol=circle,color=red,labels=[`ln(k/m)`,`ln(per)`]);
>
```

On the basis of the plot, do you think there is a power relationship between k/m and period? Place your answer below.

We now fit a line through this log-log plot; for your convenience we provide a Maple command fitLine that (1) determines the "best" line through the log-log plot of a given data set, (2) draws this line on the log-log plot of the data set, and (3) returns the equation of the line as the output of the command. Enter the following fitLine command to your logLogData.

```
> fitLine(logLogData);
>
```

Do you think this line gives a good fit to our log-log plot of periodData? Place your answer below.

Our fitLine command was set up to return the equation of the line as y=line(x). The equation for line(x) has the following form:

y = b + a*x

for some constants a and b (a=...., b=....). But what are x and y in terms of k/m and period, i.e., x=?? and y=??. This is equivalent to asking what x and y stand for in the log-log plot that you have generated above. Write out your answer on paper, then check by activating the following line.

```
> Ans(9);
```

Placing these formulas for x and y into the formula for y=line(x), we obtain an equation relating k/m and period. What is this equation? Write out your answer on paper, then check by activating the following line.

```
> Ans(10);
```

We can get rid of the extraneous logs by exponentiating both sides of this equation by base 10. Simplifying the result by the laws of logs and exponentials results in a simple formula for period in terms of k/m. Determine this formula! Write out your computations on paper, then check by activating the following line.

```
> Ans(11);
```

Referring back to the results of the fitLine command, what values have you computed for a and b? Enter them as Maple definition commands. However, round your answers for a and b to two decimal places--one of these quantities rounds to a particularly nice value. Which one? Enter your commands for a and b below.

```
> a  := ???; b  := ???;
```

We now can model period as a function of z=k/m--do this below. Complete and Enter the following function definition statement.

```
> period := z -> ???;
>
```

To check this modeling function (period as a function of z=k/m), let's plot it (an ordinary plot, not a log-log plot) along with our data periodData. The hope is that the graph of the modeling function will provide a good approximation to the scatterplot of periodData.

Go back to periodData, copy and paste it below and activate the region.

```
> periodData := ??? :
> plot1 := plot(periodData, style=POINT, symbol=CIRCLE, color=red):
> plot2 := plot(period, 5..50, color=blue):
> plots[display]([plot1, plot2], labels=[`k/m`,`period`]);
>
```

How good a match do we have? Place your answer below.

Add your equation for period as a function of k/m to the Laboratory Summary Sheet.

¤ 4. Algebraic solution.

It is useful to begin this section by removing any values that we have previously assigned to the variables m, x0, and k.

Enter the following command.

```
> m := 'm':  x0 := 'x0':  k := 'k': i:= 'i':
```

We continue with an ideal spring/mass system, i.e., with no damping force of air resistance. The IVP model for our system is:

d^2x/dt^2 + (k/m)*x = 0, where x(0) = x0, and dx/dt(0) = 0.

In the previous three subsections we generated graphical solutions for this IVP numerically, and then studied the solutions to determine the dependence of the amplitude and the period on the quantities x0 and k/m. However, as you may recall, this IVP can be solved exactly, i.e., the solution function x=x(t) can be expressed in terms of elementary functions--this is done in the subsection "A Symbolic Solution" of Section 6.2 of Calculus: Modeling and Application. The solution--which is our modeling function--is

x(t) = x0*cos(sqrt(k/m)*t).

It is a useful exercise to verify that this function does indeed satisfy the given differential equation and both of the initial conditions.

Define x(t) to be this modeling function given in terms of the (at the moment unspecified) constants m, x0, and k. Complete and Enter the following function definition.

```
> x := t -> x0*cos(sqrt(k/m)*t);
>
```

Plot this function with the set of values for k, m, and x0 used in Section 4 of the Preliminaries, i.e., k=4, m=.1, and x0=.15. (Use a standard plot command.) Enter your commands below.

```
> x0 := .15: k := 4: m:= .1:
> plot(x(t), t=0..1.8, color=green);
>
```

Does this look like the solution plot you got earlier via numerical methods? Place your answer below.

Amplitude and Period. To answer the next questions you must be able to calculate the amplitude and period of a trigonometric function of the form y = A*cos(w*t),
where A and w are constants. The amplitude equals A, and the period is that time T for which w*T=2*Pi (why?).

From the formula for x(t), determine both the amplitude and the period for x(t) as expressions in terms of x0 and k/m. Place your answers below.

Answer: amplitude = ??? period = ???

Do the formulas you have just obtained for amplitude and period exhibit the type of dependence on x0, k, and m which you predicted in Section 1 and Section 2? (In particular, if you did the computations in the optional Section 3, does your formula for the period match what you obtained in that section?) Place your answers below.

¤ 5. Empirical data.

Let's recap where we are. We have numerically analyzed the differential equation model for an ideal spring/mass system, and then followed up with an analysis of the exact solution of the model. It's time to look at some actual empirical data and apply all we have learned in our prior investigations. (Moreover, we can see if our theoretical model is a good one, i.e., if it matches the empirical data.)

For this section we will need to use the modeling function x(t) as defined in Section 4 above. Check that you do in fact have x(t) currently defined by entering the following command. If you do not get the proper output, go back to Section 4 and reenter the defining code for x(t). Enter the following command.

```
> x(t);
>
```

Our goal for this section is to examine a spring motion data set and determine values for x0, k, and m such that the graph of x(t) will give a good approximation to the scatter plot of the data.

We will determine the values for the constants by using the relationships between them and the amplitude and period of the spring motion.

• a. Defining the data.

We will use a data set HeightData1 generated by the Project CALC team at Duke University. They used an apparatus to which a spring and a mass can be attached and set into oscillation; a sensor then records the height of the mass from the bottom of the apparatus at a series of discrete times. Our data set used a mass of m=550 gm and took height measurements at time intervals of Deltat = 0.025 seconds over a total of 4 seconds. Enter mass and HeightData1.

```
> m := ???;
```

Activate the following data set HeightData1.

```
> heightData1 := [ 0.606, 0.606, 0.609, 0.616, 0.626, 0.639, 0.655, 0.674, 0.694, 0.716, 0.738, 0.
> 761, 0.783, 0.804, 0.826, 0.844, 0.857, 0.869, 0.880, 0.886, 0.888, 0.886, 0.880, 0.872, 0.858, 0
> .845, 0.828, 0.809, 0.787, 0.765, 0.743, 0.719, 0.700, 0.679, 0.661, 0.643, 0.629, 0.619, 0.612,
> 0.608, 0.608, 0.612, 0.619, 0.630, 0.643, 0.659, 0.678, 0.700, 0.722, 0.743, 0.766, 0.788, 0.808,
> 0.827, 0.845, 0.859, 0.870, 0.880, 0.884, 0.886, 0.883, 0.877, 0.868, 0.854, 0.839, 0.822, 0.804
> , 0.782, 0.760, 0.738, 0.716, 0.695, 0.676, 0.657, 0.642, 0.630, 0.620, 0.613, 0.610, 0.611, 0.61
> 5, 0.622, 0.633, 0.647, 0.663, 0.681, 0.701, 0.723, 0.747, 0.769, 0.790, 0.810, 0.829, 0.845, 0.8
> 59, 0.870, 0.878, 0.883, 0.883, 0.881, 0.874, 0.866, 0.853, 0.838, 0.821, 0.801, 0.781, 0.759, 0.
> 738, 0.716, 0.696, 0.675, 0.657, 0.643, 0.630, 0.621, 0.615, 0.613, 0.613, 0.618, 0.626, 0.637, 0
> .651, 0.667, 0.685, 0.705, 0.726, 0.747, 0.769, 0.790, 0.812, 0.830, 0.843, 0.856, 0.869, 0.877,
> 0.881, 0.881, 0.877, 0.872, 0.862, 0.849, 0.835, 0.817, 0.798, 0.778, 0.757, 0.735, 0.714, 0.695,
> 0.675, 0.659, 0.644, 0.633, 0.624, 0.619, 0.617, 0.619, 0.624, 0.632, 0.643, 0.657]:
>
```

• b. Plotting the data.

We begin our analysis of HeightData1 with a plot. However, the data does not come in ordered pairs, i.e., only the heights are given. If we plot the data set as given, the time axis will have the wrong scaling. To prevent this, we define an expanded data set TimeHeightData which contains the time as well as the height. This will plot correctly. Enter the following set of commands.

• Maple Tip. The Maple function nops returns the number of operands of an expression. In the case of a list such as heightData1, this means the number of entries in the list or the "length" of the list. Thus, for lists, you can translate nops as "length".

```
> dataSet := heightData1:
> Deltat := .025:
> timeData := [ seq( (j-1)*Deltat, j=1..nops(dataSet))]:
> timeHeightData := Zip(timeData, dataSet):
> plot(timeHeightData, color=COLOR(RGB,1,0,.3), title=`TimeHeightData`, labels=[` Time`,`H
> eight `]);
>
```

Given that we wish to compare this data plot with the graph of our modeling function x(t), we have a problem with the equilibrium height of the data: the height values are not centered on zero. In the next subsection we will rescale the height so that the equilibrium height is zero, as assumed by our modeling function x(t).

¤ c. Centering the data.

Given the need to shift the height data so that the equilibrium point is at zero, we first estimate what the current equilibrium value is. This is most easily done by computing the average (or mean) of all the height values. To do this we define a new Maple function listMean which computes the mean of any list. Enter the following Maple definition for listMean. It takes the sum of the entries divided by the number of entries.

```
> i :='i':
> listMean := proc (list) local n;
> n := nops(list):     # nops = Length
>     sum(list[i], i =1..n)/n; end:
```

We can now compute the mean of HeightData1. Enter the following command.

```
> mean := listMean(heightData1);
>
```

We can now subtract mean from each height value, obtaining a new, centered data set xData. (For convenience we will also "flip" the data across the t-axis so that we start at a maximum value rather than at a minimum.) We then plot this new data set.

• Maple Tip. The map command map(fn, list) applies the function fn to each member of the list.

Enter the following commands.

```
> xData := map(x -> -(x-mean), heightData1):
> xTimeData := Zip(timeData, xData):
> dataPlot := plot(xTimeData, color=COLOR(RGB,1,0,.3), title=`"xTimeData"`,
>     labels=[`Time`, `C- height.`] ):";
>
```

Has the desired "centering" taken place? (Yes, No)

¤ d. Estimating the constants in the model.

We now get into the heart of our analysis of the data set. Recall the form of the model function x(t) determined in Section 4:

x(t) = x0*cos(sqrt(k/m)*t), for some constants x0, k, and m.

We wish to determine values for the constants x0, k, and m so that x(t) will give a good approximation to our data. We have already assigned the value of the constant m, the mass 550. There is also a simple physical meaning for the constant x0, which enables us to determine the value of x0 directly from the data set xData. Enter your value for x0 below.

x0 := ???

We are left with determining the value of k, the spring constant. To do this, recall the computations in Section 4. In that section we found a quantity T, easily measurable from our graphs, which is inversely proportional to sqrt(k/m).

(1) What is the quantity T, and what is the equation relating T to sqrt(k/m)?

(2) Solve this equation for k, expressing k in terms of T, m, and Pi. Place your answers below.

Answers: (1) sqrt(k/m) = ?? / ???, where T equals .... (2) k = ???

Make a careful estimate of the quantity T from the graph dataPlot, and then use this to assign a value to k by typing into Maple your formula (2) from above. Complete and Enter the following commands.

T := ???
k := ???

¤ e. Graphing and refining the model function.

You should now have values defined for m, x0, and k , and a specific modeling function x(t). Be sure that you get numerical values for the first three entries.

We are now ready for the big test! We have determined values for all the constants that appear in the modeling function x(t), and so we can plot it. In particular, we can plot x(t) along with a plot of our data set. Enter the following commands.

```
> m := ???: x0 := ???:   k := ???:
> modelPlot := plot(x(t),t=0..4):";
> plots[display]([dataPlot, modelPlot],   title=`"Model & Data"`);
>
```

How good a match did you obtain? You should now refine your values of the three constants by reEntering the above command as often as necessary. Stop when you are satisfied with the results. Enter your commands above.

¤ f. The final comparison!

We now come back full circle, returning to the original differential equation model for the ideal spring/mass system. Since you now have values for all the constants in this differential equation, you can generate numerical solutions using the command idealSpring. Enter the following command.

```
> numPlot := idealSpring(k, m, x0, 500, .008):";
>
```

Does your numerically generated curve match the data plot? Let's check the match with the following (ultimate) plot. Enter the following command.

```
> plots[display]([numPlot,modelPlot, dataPlot]);
>
```

Point out any discrepancies between the three graphs that you might find, particularly in the period and amplitude of the three graphs. Might the discrepancies be "fixable" by picking better values for k and x0, or are the differences unavoidable due to the nature of our model? Are the differences serious enough to discard the function x(t) as a model for the data set under consideration? Place your answers below.

¤ 6. Extending the data set. (Optional)

As it happens, Duke took measurements on the movement of its spring/mass system for more than 4 seconds--measurements were taken for almost 8 seconds. The full data set is given below as HeightData2. We now wish to determine whether the model obtained in the previous sections for the first four seconds continues to approximate the spring/mass system over the full eight seconds. Enter the following data set.

Open this line to find HeightData2.

```
> heightData2 := [ 0.606, 0.606, 0.609, 0.616, 0.626, 0.639, 0.655, 0.674, 0.694, 0.716, 0.738, 0.
> 761, 0.783, 0.804, 0.826, 0.844, 0.857, 0.869, 0.880, 0.886, 0.888, 0.886, 0.880, 0.872, 0.858, 0
> .845, 0.828, 0.809, 0.787, 0.765, 0.743, 0.719, 0.700, 0.679, 0.661, 0.643, 0.629, 0.619, 0.612,
> 0.608, 0.608, 0.612, 0.619, 0.630, 0.643, 0.659, 0.678, 0.700, 0.722, 0.743, 0.766, 0.788, 0.808,
> 0.827, 0.845, 0.859, 0.870, 0.880, 0.884, 0.886, 0.883, 0.877, 0.868, 0.854, 0.839, 0.822, 0.804
> , 0.782, 0.760, 0.738, 0.716, 0.695, 0.676, 0.657, 0.642, 0.630, 0.620, 0.613, 0.610, 0.611, 0.61
> 5, 0.622, 0.633, 0.647, 0.663, 0.681, 0.701, 0.723, 0.747, 0.769, 0.790, 0.810, 0.829, 0.845, 0.8
> 59, 0.870, 0.878, 0.883, 0.883, 0.881, 0.874, 0.866, 0.853, 0.838, 0.821, 0.801, 0.781, 0.759, 0.
> 738, 0.716, 0.696, 0.675, 0.657, 0.643, 0.630, 0.621, 0.615, 0.613, 0.613, 0.618, 0.626, 0.637, 0
> .651, 0.667, 0.685, 0.705, 0.726, 0.747, 0.769, 0.790, 0.812, 0.830, 0.843, 0.856, 0.869, 0.877,
> 0.881, 0.881, 0.877, 0.872, 0.862, 0.849, 0.835, 0.817, 0.798, 0.778, 0.757, 0.735, 0.714, 0.695,
> 0.675, 0.659, 0.644, 0.633, 0.624, 0.619, 0.617, 0.619, 0.624, 0.632, 0.643, 0.657, 0.673, 0.692
> , 0.711, 0.732, 0.753, 0.774, 0.794, 0.814, 0.831, 0.846, 0.859, 0.869, 0.874, 0.878, 0.878, 0.87
> 4, 0.867, 0.858, 0.843, 0.827, 0.810, 0.791, 0.771, 0.750, 0.730, 0.709, 0.690, 0.671, 0.656, 0.6
> 42, 0.631, 0.623, 0.618, 0.619, 0.621, 0.626, 0.635, 0.646, 0.661, 0.677, 0.696, 0.715, 0.736, 0.
> 756, 0.777, 0.797, 0.816, 0.832, 0.847, 0.859, 0.868, 0.874, 0.877, 0.876, 0.872, 0.865, 0.855, 0
> .842, 0.827, 0.809, 0.790, 0.770, 0.750, 0.729, 0.709, 0.688, 0.670, 0.655, 0.642, 0.632, 0.624,
> 0.620, 0.619, 0.621, 0.627, 0.635, 0.647, 0.662, 0.678, 0.696, 0.716, 0.736, 0.757, 0.779, 0.799,
> 0.817, 0.833, 0.847, 0.858, 0.867, 0.873, 0.874, 0.874, 0.869, 0.862, 0.851, 0.838, 0.823, 0.806
> , 0.786, 0.767, 0.746, 0.726, 0.707, 0.688, 0.671, 0.655, 0.642, 0.632, 0.625, 0.622, 0.621, 0.62
> 4, 0.630, 0.639, 0.651, 0.666, 0.682, 0.700, 0.719, 0.739, 0.759, 0.779, 0.798, 0.817, 0.833, 0.8
> 47, 0.858, 0.865, 0.870, 0.872, 0.871, 0.866, 0.858, 0.847, 0.834, 0.819, 0.801, 0.783, 0.761, 0.
> 741, 0.722, 0.702, 0.685, 0.668, 0.654, 0.642, 0.633, 0.627, 0.624]:
```

In Section 4 we defined the Maple function x(t) to be the modeling function given in terms of the constants m, x0, and k. Make sure that x(t) remains defined in this way. Enter the following checking line.

```
> x(t);
```

We will also need the appropriate values assigned for the three constants in the following list. These were obtained in Section 5. Enter this list. Be sure that you get the correct numerical values for each entry.

```
> m; x0; k;
>
```

We wish to know whether the model obtained in the previous section continues to approximate the spring/mass system over the full time span. To determine this, we will repeat the procedures of Section 5 in order to:

¤ center the extended eight second data set;

¤ construct the two plots dataPlot (empirical data) and modelPlot (graph of the modeling function); and

¤ place both plots together in one picture for easy comparison.

The following commands center the data and produce the graph dataPlot. Enter the following commands.

```
> dataSet := heightData2:
> listMean := proc (list)
>    sum(list[i], i=1..nops(list))/nops(list); end:        # Translate  nops = length.
> mean := listMean(dataSet);
> Deltat := .025:
> timeData := [ seq( (j-1)*Deltat, j=1..nops(dataSet))]:
> xData := map(x -> -(x-mean), dataSet):
> xTimeData := Zip(timeData, xData):
> dataPlot := plot(xTimeData, color=COLOR(RGB,1,0,.3), title=`"xTimeData"`,
>    labels=[`Time`, `C- height.`] ):";
```

The following commands produce the graph modelPlot, the graph of the proposed modeling function, on top of dataPlot, the plot of the empirical data. Enter the following commands.

```
> m := m;  x0 := x0;  k := k;
> modelPlot :=plot(x(t),t=0..8):";
> plots[display]([dataPlot, modelPlot], title=`"Model & Data"`);
>
```

Do you think that the model function x(t) continues to give a good fit to the data over the expanded range of t values? If not, do you think that a good fit can be obtained by merely altering the values of x0 and/or k? Do you think we have ignored any forces in our model that perhaps are more important than we originally thought? Explain. Place your answers below.

¤ 7. Another data set.

We will briefly examine another data set HeightData3 generated by the Project CALC team at Duke University. This data set again used a mass of m=550 gm and took height measurements at time intervals of Deltat = 0.025 seconds, but this time over a total of nearly 10 seconds. Moreover, unlike the previous set of spring measurements, a piece of cardboard, parallel to the table surface, was attached to the mass. Enter HeightData3.

```
> heightData3 := [0.484, 0.487, 0.493, 0.504, 0.520, 0.541, 0.564, 0.590, 0.618, 0.649, 0.680, 0.
> 711, 0.742, 0.773, 0.801, 0.827, 0.851, 0.872, 0.888, 0.901, 0.910, 0.915, 0.915, 0.912, 0.904, 0
> .890, 0.877, 0.857, 0.836, 0.812, 0.787, 0.760, 0.732, 0.704, 0.677, 0.651, 0.625, 0.603, 0.582,
> 0.564, 0.550, 0.539, 0.532, 0.529, 0.529, 0.533, 0.541, 0.553, 0.566, 0.584, 0.603, 0.624, 0.647,
>  0.671, 0.695, 0.720, 0.746, 0.770, 0.792, 0.812, 0.829, 0.844, 0.856, 0.865, 0.870, 0.873, 0.871
> , 0.867, 0.859, 0.849, 0.836, 0.820, 0.803, 0.783, 0.764, 0.742, 0.720, 0.699, 0.675, 0.655, 0.63
> 6, 0.619, 0.604, 0.591, 0.580, 0.573, 0.569, 0.568, 0.569, 0.575, 0.582, 0.592, 0.605, 0.619, 0.6
> 35, 0.653, 0.671, 0.691, 0.710, 0.729, 0.747, 0.765, 0.782, 0.796, 0.809, 0.819, 0.828, 0.833, 0.
> 836, 0.836, 0.833, 0.828, 0.822, 0.812, 0.800, 0.787, 0.772, 0.756, 0.739, 0.722, 0.704, 0.687, 0
> .671, 0.656, 0.642, 0.629, 0.617, 0.610, 0.602, 0.600, 0.598, 0.598, 0.601, 0.606, 0.614, 0.624,
> 0.635, 0.648, 0.662, 0.677, 0.692, 0.707, 0.723, 0.738, 0.753, 0.766, 0.778, 0.788, 0.797, 0.804,
>  0.809, 0.811, 0.812, 0.810, 0.806, 0.800, 0.792, 0.783, 0.772, 0.760, 0.747, 0.733, 0.720, 0.705
> , 0.691, 0.678, 0.665, 0.654, 0.643, 0.635, 0.627, 0.622, 0.618, 0.617, 0.617, 0.620, 0.625, 0.63
> 0, 0.638, 0.649, 0.660, 0.671, 0.683, 0.694, 0.709, 0.722, 0.734, 0.746, 0.758, 0.768, 0.777, 0.7
> 84, 0.788, 0.792, 0.794, 0.794, 0.793, 0.790, 0.785, 0.778, 0.770, 0.762, 0.751, 0.741, 0.729, 0.
> 718, 0.706, 0.694, 0.683, 0.672, 0.662, 0.654, 0.647, 0.640, 0.636, 0.633, 0.632, 0.632, 0.635, 0
> .638, 0.644, 0.650, 0.658, 0.667, 0.677, 0.687, 0.698, 0.710, 0.721, 0.732, 0.742, 0.751, 0.760,
> 0.767, 0.774, 0.778, 0.782, 0.783, 0.783, 0.782, 0.779, 0.775, 0.769, 0.763, 0.755, 0.746, 0.737,
>  0.725, 0.715, 0.705, 0.696, 0.686, 0.677, 0.669, 0.662, 0.656, 0.651, 0.647, 0.645, 0.644, 0.645
> , 0.647, 0.650, 0.655, 0.661, 0.667, 0.675, 0.683, 0.692, 0.701, 0.710, 0.719, 0.728, 0.737, 0.74
> 5, 0.752, 0.758, 0.763, 0.768, 0.771, 0.772, 0.773, 0.771, 0.769, 0.765, 0.760, 0.754, 0.747, 0.7
> 40, 0.732, 0.724, 0.715, 0.707, 0.698, 0.690, 0.682, 0.676, 0.670, 0.663, 0.659, 0.656, 0.654, 0.
> 654, 0.654, 0.656, 0.659, 0.664, 0.669, 0.675, 0.681, 0.689, 0.696, 0.704, 0.712, 0.720, 0.728, 0
> .735, 0.741, 0.747, 0.752, 0.757, 0.760, 0.762, 0.762, 0.762, 0.761, 0.758, 0.755, 0.750, 0.745,
> 0.739, 0.732, 0.725, 0.718, 0.711, 0.703, 0.698, 0.691, 0.685, 0.679, 0.674, 0.670, 0.666, 0.665,
>  0.664, 0.663, 0.664, 0.666, 0.669, 0.673, 0.678, 0.683, 0.689, 0.694, 0.702, 0.707, 0.714, 0.721
> , 0.728, 0.734, 0.740, 0.744, 0.749, 0.752, 0.754, 0.756, 0.756, 0.755, 0.754, 0.751, 0.748, 0.74
> 4, 0.739, 0.733, 0.728, 0.721, 0.715, 0.708, 0.702, 0.696, 0.690, 0.684, 0.680, 0.675, 0.672, 0.6
> 70, 0.668, 0.667, 0.667, 0.669, 0.671, 0.674, 0.677, 0.684, 0.689]:
```

As done for all the previous data sets, we use the following set of commands to produce a plot of the oscillations of the spring/mass system as centered on the equilibrium position. Enter the following set of commands.

```
> dataSet := heightData3:
> listMean := proc (list)
>    sum(list[i], i=1..nops(list))/nops(list); end:          # Translate  nops = length.
> mean := listMean(dataSet);
> Deltat := .025:
> timeData := [ seq( (j-1)*Deltat, j=1..nops(dataSet))]:
> xData := map(x -> -(x-mean), dataSet):
> xTimeData := Zip(timeData, xData):
> dataPlot := plot(xTimeData, color=COLOR(RGB,1,0,.3), title=`"xTimeData"`,
>    labels=[`Time`, `C- height.`] ):";
```

Is the model of the ideal spring a good model for this particular spring oscillation, i.e., do you think there exist values for the constants x0, k, and m such that the graph of the modeling function x(t) from Section 4 would well approximate the above scatter plot? Explain what you see in the scatter plot that motivates your answer. Place your answer below.

What force have we ignored in constructing our model that is, in fact, an important factor in this particular spring oscillation? Why would this force affect the scatter plot in the way seen above? Place your answer below.

Examining the above scatter plot should lead you to guess that the form of a good modeling function would be the product of two familiar types of elementary functions. What elementary functions could be used in this way? Place your answer below.

The data set HeightData3, considered in this section, is examined in great depth in the lab entitled "Damped Spring Motion." In particular, a more sophisticated model than the "ideal spring" IVP is developed and studied in that laboratory.

□ LABORATORY SUMMARY SHEET.

¤ Varying initial displacements x0.

k/m = 40; x0 varies.

x0	amplitude	period
______	______	______
______	______	______
______	______	______
______	______	______
______	______	______
______	______	______

Conclusions. How does x0 affect amplitude?
How does x0 affect the period?

¤ Varying ratios k/m.

k/m varies; x0 = .15

k/m	amplitude	period
______	______	______
______	______	______
______	______	______
______	______	______
______	______	______
______	______	______

Conclusions:
How does k/m affect amplitude?

How does k/m affect the period?

(Optional) Give a formula for period as a function of k/m. □

LAB 12C: Damped Spring Motion

□ NAMES:

□ PURPOSE.
As described in Calculus: Modeling and Application, the oscillation of an ideal spring/mass system depends only on the ratio of the mass and the spring constant, along with the starting displacement. In this lab we extend the model to account for damping by a retarding frictional force. In particular, we will focus on determining the relationship between all these quantities and the period and amplitude of the solution curves. These relationships will be studied from three different viewpoints: numerical, algebraic (symbolic), and empirical.

□ PREPARATION.
Prior to the laboratory session you should:

Read Sections 6.1 and 6.2 of Calculus: Modeling and Application.

Read the Preliminaries, writing out answers to the questions which are asked. During the lab session you will then be able to check your work against the closed answer cells in the worksheet and move quickly to the Project.

□ LABORATORY REPORT INSTRUCTIONS.
For this laboratory you are not required to submit a formal report. Rather you are asked to submit

• An electronic copy of this Worksheet with the questions answered in the indicated locations. Only one submission will be made. Write in complete sentences and be sure the sentences connect together in coherent ways.

• A filled-in paper copy of the Laboratory Summary Sheet. This is the last section of the laboratory worksheet.

□ Maple Procedures.

Be sure to initialize the Worksheet before beginning the lab.

```
> with(lab12c);
> # If an Error appears as an output for the above line, see your lab instructor.
```

Lab 12C Procedures define a command that will be a central tool for this lab: dampedSpring. This command generates numerical solutions for any damped spring/mass system. This commands will be further explained when needed.

□ PRELIMINARIES.

¤ 1. A Differential Equation Model for a Damped Spring/Mass System.

In this lab we consider the movement of a mass on a spring, as described in Chapter 6 of Calculus: Modeling and Application. We will, however, make the model more realistic by taking account of the damping effect of air resistance. The following six steps lead you through the derivation of the initial value problem which models such a system. This parallels the discussion in the subsections "The Spring-Mass System" and "The Initial Value Problem" of Section 6.2 of Calculus: Modeling and Application, differing from the text only by the inclusion of the force of air resistance.

What follows is a rendering of Figure 6.9, page 367 of Calculus: Modeling and Application.

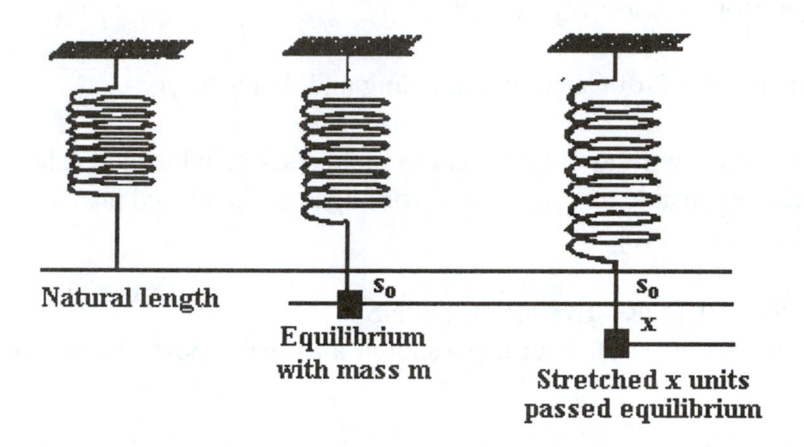

¤ Step 1. Suppose a mass m stretches a spring by s0 units from its natural length. Then, according to Hooke's Law, the force exerted by the spring on the mass is -k*s0, where k is the spring constant (we will denote downward forces as positive, and upward forces as negative). However, the force of gravity, m*g, will balance the spring force, so that 0 = ..., which gives .....=...... Check your answer by activating the following line.

```
> Ans(1);
```

¤ Step 2. Take the spring mass system from Step 1, stretch it beyond the equilibrium s0 that it had achieved in Step 1, and then let go. At any time t let x(t) be the distance beyond the equilibrium s0 to which the mass is pulled (downward displacement is positive). There will be three significant forces acting on the mass. Write answers to the questions on paper, then activating the command line which follows.

1. The force of gravity = ..., which from Step 1 can be alternately expressed as ......
2. The force of the spring = ...
3. The force of air resistance = ... (assumed to be proportional to velocity).

```
> Ans(2);
```

¤ Step 3. Thus, at time t, the total force on the mass will be F = ... Check your answer by activating the following line.

```
> Ans(3);
```

¤ Step 4. By Newton's second Law of Motion, a force F on a mass m will cause an acceleration a, where the three quantities are related by the formula F = ... Moreover, the acceleration can be expressed in terms of the position x=x(t) of the mass as a = ... Check your answers by activating the following line.

```
> Ans(4);
```

¤ Step 5. Combining the results of Steps 3 and 4 yields a differential equation model for the motion of a spring with damping. How does this compare with the equation 367 of Calculus: Modeling and Application? Write down the equation on paper, then activate the following line.

```
> Ans(5);
```

¤ Step 6. Assume that initially (t=0) the spring is stretched a distance x0 beyond the equilibrium it had achieved in Step 1, and is then released. This gives two initial conditions that must be satisfied by x(t):

x(0) = ..., and dx/dt(0) = ... Check your answers by activating the following line.

```
> Ans(6);
```

The differential equation of Step 5, combined with the initial conditions of Step 6, gives a second-order initial value problem (IVP) model for the motion of a damped spring-mass system.

¤ 2. Euler's Method for the Damped Spring/Mass Model.

Using the same methods as employed in Section 6.2, starting on page 368, of Calculus: Modeling and Application for an undamped spring, we convert our second-order IVP model into a system of two coupled first-order IVPs. To do this we introduce the velocity function v(t) = dx/dt. Then x and v must satisfy the following system of IVPs:

dx/dt = v, x(0) = x0
dv/dt = - (a/m)*v - (k/m)*x, v(0) = 0

Do you see how these equations follow from the IVP given in Steps 5 and 6 of the previous section?

Recall that x(t) is the (downward) displacement from the system's equilibrium position at time t,
v(t) is the velocity of the mass at time t, i.e., the derivative of x(t),
x0 is the initial displacement of the mass,
m is the mass at the end of the spring,
k is the spring constant from Hooke's Law,
a is the system's proportionality constant for air resistance in terms of velocity.

We will begin the Project by approximating solutions to this "double" initial value problem by using Euler's Method simultaneously on the two initial value problems for x and v--we must work on both equations simultaneously because the variables x and v are intertwined. (This is the vector Euler's Method, first considered in the lab on epidemics.)

Starting with specific values for x0, m, k, a, and Deltat (the time step size), we will apply the Euler Method between 500 and 1000 times; the values of the quantities x and v so produced at time t=k*Deltat will be denoted by x(k) and v(k). We will repeat this procedure for various values of a in order to determine the effects of air resistance which are predicted by our spring/mass model.

¤ 3. Deriving the "Updating Equations" for Euler's Method.

Given the values x(k) and v(k), we wish to derive equations for the next set of values x(k+1) and v(k+1). To do this we use Euler's Method, the central idea being the replacement of derivatives with difference quotients in the two differential equations from the previous section. At the k-th step define

Deltax = x(k+1)-x(k) and Deltav = v(k+1)-v(k).

Replace the derivatives with difference quotients (Deltax /Deltat and Deltav/Deltat) in the two differential equations from section 1. Write your results on paper, then check them by activating the following line.

```
> Ans(7);
```

Solve these two equations for x(k+1) and v(k+1). Write your results on paper, then check them by activating the following line.

```
> Ans(8);
>
```

We have the starting values x(0) and v(0)--these are simply the initial conditions. Now, however, we can use the equations just derived to obtain the first set of computed values, x(1) and v(1), and from these to obtain the second set of computed values, x(2) and v(2), and so on. This is how Euler's Method will be implemented in the next section.

¤ 4. The Maple Commands--Implementing Euler's Method.

For various values of the constants a, k, m, and x0 we wish to use Euler's Method to obtain numerical solutions for the damped spring/mass model. To do this conveniently, we have constructed the Maple command DampedSpring. Find out about this new command by the use of ?name. Enter your commands below.

As an example of the this command, the statements below will generate a plot of displacement verses time for a damped spring/mass system with a=.1, k=4, m=.1, and x0=.15, using n=600 steps with a step size of Deltat=.005. Enter the following region and examine the output.

```
> Alpha :=.1: k :=4: m :=.1: x0 :=.15: n :=600: Deltat :=.005:
> dampedSpring(Alpha, k, m, x0, n, Deltat);
>
```

The graph of the displacement x=x(t) generated above looks very much like the graph of the product of two well-known types of functions. Can you guess what these functions are? Place your guess below.

□ PROJECT.

¤ 1. Modest damping--Numerical solution.

We consider a spring/mass system for which there is a modest damping force of air resistance. This is the situation considered in Preliminaries, and corresponds to the case a > 0 (with a not too large). From the development in Section 1 of the Preliminaries, the IVP model for our system is:

d^2x/dt^2 + (a /m)*(dx/dt) + (k/m)*x = 0,

where x(0) = x0, and dx/dt(0) = 0.
We will study this model numerically by using (a "super-charged") Euler's Method to generate numerical solutions for a variety of values of a, k, m, and x0. We will take special notice of the effect of a on the amplitude and period of the solution curves.

Let's begin with the example used in Section 4 of the Preliminaries. We repeat below the relevant Maple commands from that section. Enter the following commands.

```
> Alpha :=.1: k :=4: m :=.1: x0 :=.15: n :=600: Deltat :=.005:
> dampedSpring(Alpha, k, m, x0, n, Deltat,Euler=false);
>
```

How does the behavior of the amplitude differ in this damped case as opposed to the undamped case (the "ideal spring") which you studied in Calculus: Modeling and Application? Is this new behavior physically realistic? Place your answer below.

In order to crudely measure the property you probably identified in answer to the previous question, we define the amplitude decline as the ratio of the height of the first peak divided by the initial amplitude--this will indicate how rapidly the amplitude has declined during the first full cycle. Compute the amplitude decline for the curve given above. Enter your computations and answer below.

As in the case of the "ideal spring", is there still a constant period for the damped motion, i.e., is the length of time required to make a complete "cycle" the same no matter where you begin on the curve? (Hint: If this is so, then the zeros in our graph will all be equidistant from each other.) Enter your computations and answers below.

• Maple Tip. Remember that you can "pick" points off of a plot by clicking the left mouse button and the coordinates appear on the lower left boarder of the plot.

Determine the initial amplitude, the amplitude decline, and the period of this solution curve, and record your answers in the table below, as well as in your Laboratory Summary Sheet. Place your answers below.

Alpha	initial amplitude	amplitude decline	period
???	???	???	???
???	???	???	???
???	???	???	???
???	???	???	???

```
> Alpha :=.3:  k :=4:  m :=.1:  x0 :=.15:  n :=600:  Deltat :=.005:
> dampedSpring(Alpha, k, m, x0, n, Deltat);
```

Now change the value of the damping constant Alpha, rerun the DampedSpring commands from above with this new value, and record the initial amplitude, the amplitude decline, and the period into the table above and into your Laboratory Summary Sheet. Repeat this for enough values of a so that you see the consequence (or non-consequence) of changes in Alpha on the recorded quantities. Record your answers into the table and the Laboratory Summary Sheet.

In what ways does the value of Alpha affect the amplitude of the spring oscillation? In particular, can you make the amplitude decline ratio as small as you desired by adjusting the value of Alpha? Is your answer surprising, and does it make physical sense? Place your answer below and in the Laboratory Summary Sheet.

In what way does the value of Alpha affect the period of the spring oscillation? In particular, can you make the period as large as you desired by adjusting the value of Alpha? Does your answer make physical sense? Place your answer below and in the Laboratory Summary Sheet.

It would be interesting to try and determine more precisely the nature of the relationships between Alpha and the amplitude and period via numerical methods. Equations for these relationships can be obtained numerically--however, time does not allow us to pursue this method further. Moreover, in the next section we shall determine the desired formulas by a different method: we can find exact algebraic solutions for the damped spring/mass IVP.

¤ 2. Modest damping--Algebraic solution.

We continue with a spring/mass system for which there is a significant damping force of air resistance (Alpha > 0). Recall that the IVP model for our system is:

d^2x/dt^2 + (Alpha/m)*(dx/dt) + (k/m)*x = 0,

where x(0) = x0, and dx/dt(0) = 0.

In the previous section we generated solutions for this IVP numerically, and then studied the solutions to determine the dependence of the amplitude and the period on the damping constant Alpha. However, as in the case of the undamped spring which you studied in Calculus: Modeling and Application, this IVP can be solved exactly, i.e., the solution function x=x(t) can be expressed in terms of elementary functions.

Although this IVP may look formidable to you as a beginner in calculus, it is, in fact, a member of a well understood collection of differential equations ("2nd order, homogeneous, linear, constant coefficient") for which an algorithm for obtaining the solutions is well known. This algorithm is a standard topic in any differential equations course, and (for time considerations) will not be described in this lab.

A less sophisticated (but still effective) method for solving the IVP is to guess at the general form of a solution function (containing a sufficient number of as-yet-to-be-determined constants), then algebraically attempt to find values for the constants which produce an explicit solution. In fact, in Section 4 of the Preliminaries, we asked you to make such a guess at the general form of the solution curves: we were hoping that you might conjecture that each solution is the product of a decaying exponential times a cosine function. In fact, this is correct. The solution to our IVP--when Alpha is not too large--is of the form

x(t) = A*e^(-b*t)*cos(w*t + d), for some constants A, b, w, and d.

In the next (optional) section, we give you a tour of the symbolic computational power of Maple by solving for the constants A, b, w, and d in terms of a, k, m, and x0. In this section we will merely state these equations and consider their consequences.

b = Alpha/(2*m)
w = sqrt(k/m - b^2)
d = - arctan(b/w)
A = x0*sec(d)

To convince you that these equations do work as claimed, let us take the specific example (i.e., specific values for Alpha, k, m, and x0) first used in the Preliminaries, and graph the function x=x(t) which comes from the equations we have just stated. First we define x(t) as a Maple function, and then we execute the equations defining A, b, w, and d in terms of Alpha, k, m, and x0. Enter the following commands.

```
> Alpha := 'Alpha':  k := 'k':  m := 'm':
> x := t -> A*exp(-beta*t)*cos(omega*t + delta);
> beta  :=  Alpha/(2*m);
> omega :=  sqrt((k/m)-beta^2);
> delta := - arctan(beta/omega);
> A     :=  x0*sec(delta);
```

Now obtain the values used earlier for Alpha, k, m, and x0 and place them below. The code will then produce a pleasing graph of the function x(t) that we have defined above. Does the plot of x(t) look like the solution plots we obtained earlier through numerical methods? Complete and Enter the following commands.

```
> Alpha := ??:  k := ??:  m := ??:  x0 := ??: n := 600: Deltat := .005:
> xPlot:=plot(x(t), t=0.. n*Deltat, y=-A..A, color=COLOR(RGB,1,0,.3)):";
```

To see how well the plot of x(t) matches our earlier numerically generated curves, let's put both of them in the same plot and compare. The following two commands accomplish this goal. Enter the following two commands.

```
> numPlot := dampedSpring(Alpha, k, m, x0, n, Deltat):";

> plots[display]([numPlot,xPlot], title=`Green=numerical, Red=algebraic`);
```

How good a match did you obtain? Place your answer below.

Now we again consider the amplitude and period of the spring motion modeled by these equations. Which term in the definition of x=x(t) given above is the cause of the decay of the amplitude? Place your answer below.

How does an increase in the damping constant Alpha affect this amplitude decay term? Does this match with the relationship observed between Alpha and amplitude decline in the previous subsection? Have we increased our knowledge about the amplitude decline? Place your answer below.

We now turn to a consideration of the period of the oscillation. Which term in the definition of x = x(t) given above is the cause of the oscillation? Place your answer below.

Give an equation which expresses the period T of the oscillation in terms of omega. Then explain how increasing the damping constant Alpha will affect the period via your equation. Does this match with the relationship observed between Alpha and period in the previous subsection? Have we increased our knowledge about this relationship? Place your answer below.

What is the largest value of Alpha for which our equation for omega (and hence our equation for the period) can still be valid? Denote this important value by Alpha0. Place your formula for Alpha0 below.

```
> Alpha0 := ???;
>
```

We define "modest damping" to be 0 < Alpha < Alpha0, while "extreme damping" will mean Alpha >= Alpha0. Use the command dampedSpring to numerically plot an example of what happens in the case of extreme damping. Describe the major differences that you see between the two cases. Enter your commands and answers below.

```
> Alpha := ???: k := 4: m := .1: x0 := .15: n := 600: Deltat := .005:
> dampedSpring(Alpha, k, m, x0, n, Deltat);
```

What happens to the period T as a gets close to, but does not exceed, Alpha0? In particular, recall the question concerning the period from the previous subsection: Can you make the period as large as you desired by adjusting the value of Alpha? (Hint: Express T in terms of omega, then omega in terms of Alpha. Plotting T as a function of Alpha is also helpful.) Place your answers below.

¤ 3. Using Maple to find the constants. (Optional)

In the last section we used the constants

b = Alpha/(2*m)
w = sqrt(k/m - b^2)
d = - arctan(b/w)
A = x0*sec(d)

to go along with the algebraic solution

x(t) = A*e^(-b*t)*cos(w*t + d),

of our initial value problem.

We continue with a spring/mass system for which there is a significant damping force of air resistance (Alpha > 0). Recall that the IVP model for our system is:

d^2x/dt^2 + (Alpha/m)*(dx/dt) + (k/m)*x = 0,

where x(0) = x0, and (dx/dt)(0) = 0.

We guess that--for Alpha small--the solution of this IVP is of the following form:

x(t) = A*e(-b*t)*cos(w*t + d), for some constants A, b, w, and d.

The objective of this subsection is to use the symbolic computation power of Maple to determine formulas for the constants A, b, w, and d in terms of a=Alpha, k, m, and x0 so that the function x(t) will be the solution of the IVP. (These formulas were stated in the previous subsection but were not derived.) This is an involved computation, so prepare to work hard! In particular, there will be a lot of variables and constants--a pencil and paper may be handy to keep track of them all.

We first define x(t) as a Maple function. Enter the following definition for x(t).

```
> A := 'A': m := 'm': k := 'k': w := 'w':
> x:= t -> A*exp(-b*t)*cos(w*t + d);
```

Use the Maple D operator to take the derivative of a function. Re-label D(x) as Dx.

```
> Dx := D(x);
> DDx := D(D(x));
```

We now examine the derivative expression that we wish to have equal to zero. Enter the following command defining expression1.

```
> expression1 := DDx(t) + (a/m)*Dx(t) + (k/m)*x(t);
```

As you can see, Maple is quite capable of computing complicated derivatives. We now "simplify" this expression to give expression2. Enter the following command.

```
> expression2 := simplify (expression1);
```

• Maple explanation. Using the simplify command causes Maple to apply a set of algebraic rules in an attempt to simplify the expression. This may or may not work, depending on the nature of the expression.

Our desire is to set expression2 equal to 0 (since expression2 is the combination of derivatives of x(t) which our model says equals 0) and see what restrictions that places on the constants A, b, w, and d. However, notice that expression2 is of the form A*e^(-b*t)*(...)/m, so that the terms A*e^(-b*t) and m can be eliminated when expression2 is set equal to zero. For that reason we will get rid of them now. Enter the following command.

```
> expression3 := simplify(exp(b*t)*(1/A)*expression2*m);
```

This expression can be rearranged into the form

c1*cos(d+t*w) + c2*sin(d+t*w), where c1 and c2 are constants.

If such an expression equals zero for all values of t, then the two constants c1 and c2 must equal zero. We find the value for c1 by effectively substituting the value of t which gives w*t+d = 0 or cos(w*t+d)=1 and sin(w*t+d)=0. Next we will compute c2 by interchanging those values.

Enter the following command to determine c1.

```
> c1 := subs(cos(w*t+d)=1, sin(w*t+d)=0, expression3);
```

• Maple explanation. We are using substitution rules. The subs command tells Maple what substitutions are to be made. The first argument says to replace cos(w*t+d) with 1, and the second says to replace sin(w*t+d) with 0. What is left after such replacements is merely the coefficient of cos(d + t*w) in expression3. It is crucially important to understand that replacement rules do not change the expression to which they are applied--they produce a new expression. Hence our command above does not change expression3--it merely produces a new expression, c1.

Find c2 by activating the following command.

```
> c2 := subs(cos(w*t+d)=0, sin(w*t+d)=1, expression3);
```

We now set c2 equal to zero and solve for b. Enter the following command.

```
> bb :=solve(c2=0, b);
>
```

• Maple explanation. Equations, as opposed to assignments, are denoted with the equal signs (=) without the colon. The solve command attempts to solve the given equation (c2 = 0) for the given variable (b). As can be seen, the result comes back in a very convenient form. Did you get what you expected? Go back to the beginning of the section and check it.

We thus have succeeded in getting a formula for b, written as a replacement rule denoted as bb. Now we set c1 equal to zero and solve for w. Enter the following command.

```
> ww1 :=[ solve(c1=0, w)];
```

Maple Tip. We place square brackets around the command so that the two solutions will be presented as a list. This will enable us to conveniently take on of them. Since we will only need one of them, we pick the second (positive) one.
Enter the following command.

```
> ww2 := ww1[2];
```

We almost have our final formula for w. However, the expression still has b in it, which is not desired. No problem--we use the subs command again to get rid of b. Enter the following command.

```
> ww := subs(b=bb, ww2);
```

Go back to the beginning of this section and verify that this is the desired result.
Our formulas for b and w (in the form of replacement rules bb and ww) should, when placed into the derivative expression in our IVP (expression1), produce 0. Let's check that this is in fact true. Enter the following two commands.

```
> s := subs(b=bb, w=ww, expression1);
> simplify(s);
```

The result of the first command should have been disturbing...but the result of the second command should have been reassuring.

We have two last constants to solve for: d and A. We obtain replacement rules for these quantities by using the two initial conditions of our IVP: x(0)=x0 and Dx(0)=0. Let us compute both x(0) and Dx(0). Enter the following commands.

```
> x(0);
> Dx(0);
```

We first solve x(0)=x0 for A. Enter the following region.

```
> x0 := .15:
> AA := solve(x(0)=x0, A);
```

This gives us an expression for A, although it is in terms of d. No problem, solve for it.
Enter the following command.

```
> dd := solve(Dx(0) = 0, d);
>
```

This gives us d in terms of b and w, and we already have formulas for b and w. Thus we have obtained all of the desired equations.

¤ 4. Empirical data.

Let's recap where we are. We have numerically analyzed the differential equation model for a damped spring/mass system, and then followed up with an analysis of the exact solution of the model. It's time to look at some actual data and apply all we have learned in our prior investigations. (Moreover, we can see if our theoretical model is a good one, i.e., if it matches real world data.)

¤ a. Defining the data.

We will use a data set heightData3 generated by the Project CALC team at Duke University. They used an apparatus to which a spring and a mass can be attached and set into oscillation; a sensor then records the height of the mass from the bottom of the apparatus at a series of discrete times. Our data set used a mass of m=550 gm and took height measurements at time intervals of Deltat = 0.025 seconds over a total of nearly 10 seconds. To ensure a significant damping factor, a piece of cardboard, parallel to the table surface, was attached to the mass. Enter mass and heightData3.

```
> m := ???;

> heightData3 := [0.484, 0.487, 0.493, 0.504, 0.520, 0.541, 0.564, 0.590, 0.618, 0.649, 0.680, 0.7
> 11, 0.742, 0.773, 0.801, 0.827, 0.851, 0.872, 0.888, 0.901, 0.910, 0.915, 0.915, 0.912, 0.904, 0.
> 890, 0.877, 0.857, 0.836, 0.812, 0.787, 0.760, 0.732, 0.704, 0.677, 0.651, 0.625, 0.603, 0.582, 0
> .564, 0.550, 0.539, 0.532, 0.529, 0.529, 0.533, 0.541, 0.553, 0.566, 0.584, 0.603, 0.624, 0.647,
> 0.671, 0.695, 0.720, 0.746, 0.770, 0.792, 0.812, 0.829, 0.844, 0.856, 0.865, 0.870, 0.873, 0.871,
>  0.867, 0.859, 0.849, 0.836, 0.820, 0.803, 0.783, 0.764, 0.742, 0.720, 0.699, 0.675, 0.655, 0.636
> , 0.619, 0.604, 0.591, 0.580, 0.573, 0.569, 0.568, 0.569, 0.575, 0.582, 0.592, 0.605, 0.619, 0.63
> 5, 0.653, 0.671, 0.691, 0.710, 0.729, 0.747, 0.765, 0.782, 0.796, 0.809, 0.819, 0.828, 0.833, 0.8
> 36, 0.836, 0.833, 0.828, 0.822, 0.812, 0.800, 0.787, 0.772, 0.756, 0.739, 0.722, 0.704, 0.687, 0.
> 671, 0.656, 0.642, 0.629, 0.617, 0.610, 0.602, 0.600, 0.598, 0.598, 0.601, 0.606, 0.614, 0.624, 0
> .635, 0.648, 0.662, 0.677, 0.692, 0.707, 0.723, 0.738, 0.753, 0.766, 0.778, 0.788, 0.797, 0.804,
> 0.809, 0.811, 0.812, 0.810, 0.806, 0.800, 0.792, 0.783, 0.772, 0.760, 0.747, 0.733, 0.720, 0.705,
>  0.691, 0.678, 0.665, 0.654, 0.643, 0.635, 0.627, 0.622, 0.618, 0.617, 0.617, 0.620, 0.625, 0.630
> , 0.638, 0.649, 0.660, 0.671, 0.683, 0.694, 0.709, 0.722, 0.734, 0.746, 0.758, 0.768, 0.777, 0.78
> 4, 0.788, 0.792, 0.794, 0.794, 0.793, 0.790, 0.785, 0.778, 0.770, 0.762, 0.751, 0.741, 0.729, 0.7
> 18, 0.706, 0.694, 0.683, 0.672, 0.662, 0.654, 0.647, 0.640, 0.636, 0.633, 0.632, 0.632, 0.635, 0.
> 638, 0.644, 0.650, 0.658, 0.667, 0.677, 0.687, 0.698, 0.710, 0.721, 0.732, 0.742, 0.751, 0.760, 0
> .767, 0.774, 0.778, 0.782, 0.783, 0.783, 0.782, 0.779, 0.775, 0.769, 0.763, 0.755, 0.746, 0.737,
> 0.725, 0.715, 0.705, 0.696, 0.686, 0.677, 0.669, 0.662, 0.656, 0.651, 0.647, 0.645, 0.644, 0.645,
>  0.647, 0.650, 0.655, 0.661, 0.667, 0.675, 0.683, 0.692, 0.701, 0.710, 0.719, 0.728, 0.737, 0.745
> , 0.752, 0.758, 0.763, 0.768, 0.771, 0.772, 0.773, 0.771, 0.769, 0.765, 0.760, 0.754, 0.747, 0.74
> 0, 0.732, 0.724, 0.715, 0.707, 0.698, 0.690, 0.682, 0.676, 0.670, 0.663, 0.659, 0.656, 0.654, 0.6
> 54, 0.654, 0.656, 0.659, 0.664, 0.669, 0.675, 0.681, 0.689, 0.696, 0.704, 0.712, 0.720, 0.728, 0.
> 735, 0.741, 0.747, 0.752, 0.757, 0.760, 0.762, 0.762, 0.762, 0.761, 0.758, 0.755, 0.750, 0.745, 0
> .739, 0.732, 0.725, 0.718, 0.711, 0.703, 0.698, 0.691, 0.685, 0.679, 0.674, 0.670, 0.666, 0.665,
> 0.664, 0.663, 0.664, 0.666, 0.669, 0.673, 0.678, 0.683, 0.689, 0.694, 0.702, 0.707, 0.714, 0.721,
>  0.728, 0.734, 0.740, 0.744, 0.749, 0.752, 0.754, 0.756, 0.756, 0.755, 0.754, 0.751, 0.748, 0.744
> , 0.739, 0.733, 0.728, 0.721, 0.715, 0.708, 0.702, 0.696, 0.690, 0.684, 0.680, 0.675, 0.672, 0.67
> 0, 0.668, 0.667, 0.667, 0.669, 0.671, 0.674, 0.677, 0.684, 0.689]:
>
```

¤ b. Plotting the data.

We begin our analysis of HeightData3 with a ListPlot. However, the data does not come in ordered pairs, i.e., only the heights are given. If we ListPlot the data set as given, the time axis will have the wrong scaling. To prevent this, we define an expanded data set TimeHeightData which contains the time as well as the height. This will ListPlot correctly. Enter the following commands.

```
> dataSet := heightData3:
> Deltat := .025:
> timeData := [ seq( (j-1)*Deltat, j=1..nops(dataSet))]:
> timeHeightData := Zip(timeData, dataSet):
> plot(timeHeightData, color=COLOR(RGB,1,0,.3), title=`TimeHeightData`, labels=[` Time`,`H
> eight `]);
```

Does there appear to be a significant amount of damping in the data? Place your answer below.

Given that we wish to compare this data plot with the model developed earlier, do you see any problems with the scaling of the dependent (height) axis? Place your answer below.

¤ c. Centering the data.

Given the need to shift the height data so that the equilibrium point is at zero, we first estimate what the current equilibrium value is. This is most easily done by computing the average (or mean) of all the height values. To do this we define a new Maple function listMean which computes the mean of any list. Enter the following Maple definition for listMean. It takes the sum of the entries divided by the number of entries.

```
> i :='i':
> listMean := proc (list)
>     sum(list[i],i =1..nops(list))/nops(list); end;
```

We can now compute the mean of heightData3. Enter the following command.

```
> mean := listMean(???);
```

We now can subtract mean from each of the heights, obtaining a new, centered data set xData. (For convenience we will also "flip" the data across the t-axis so that we start at a maximum value rather than at a minimum.) We then plot this new data set. Enter the following commands.

```
> xData := map(x -> -(x-mean), heightData3):
> xTimeData := Zip(timeData, xData):
> dataPlot := plot(xTimeData, color=COLOR(RGB,1,0,.3), title=`"xTimeData"`,
>    labels=[`Time`, `CentHt.`] ):";
```

Has the desired "centering" taken place? (Yes, No)

¤ d. Estimating the frequency: w.

We now get into the heart of our analysis of the data set. Recall the form of the model function x(t) determined in Section 2:

x(t) = A*e^(-b*t)*cos(w*t+d), for some constants A, b, w, and d.

We wish to determine values for the constants A, b, w, and d so that x(t) will give a good approximation to our data. We could try a simple-minded brute force approach: take a guess at the values of the constants, plot the graph of the corresponding x(t) on top of a plot of our data set, and then adjust the constants to achieve a good fit. The problem with this method is that with four constants to vary, it will be difficult to achieve a good fit within a reasonable amount of time.

However, from our previous work in this lab, we know how the constants are related to quantities which are measurable from the data set. Let's start with w. There is a quantity, easily measurable from the graph of our data, which is inversely proportional to w. What quantity are we referring to, and what is the equation relating this quantity to w? (Hint: See Section 2.) Place your answers below.

Using the plot dataPlot constructed above, make a careful estimate of the quantity just referred to, and then use this to assign a value to w . Enter your calculations below.

```
> w := ???;
```

¤ e. Approximating the damping envelope function x=A exp(-b*t).

If you remove the cosine term from our modeling function x=x(t), you are left with a decaying exponential of the form x = A*exp(-b*t). This is called the damping envelope function, since it "envelopes" the graph of the data by sliding along the oscillation peaks. As an example, the blue curve shown in the next plot is the graph of the damping envelope function for the oscillating red curve shown beneath it.

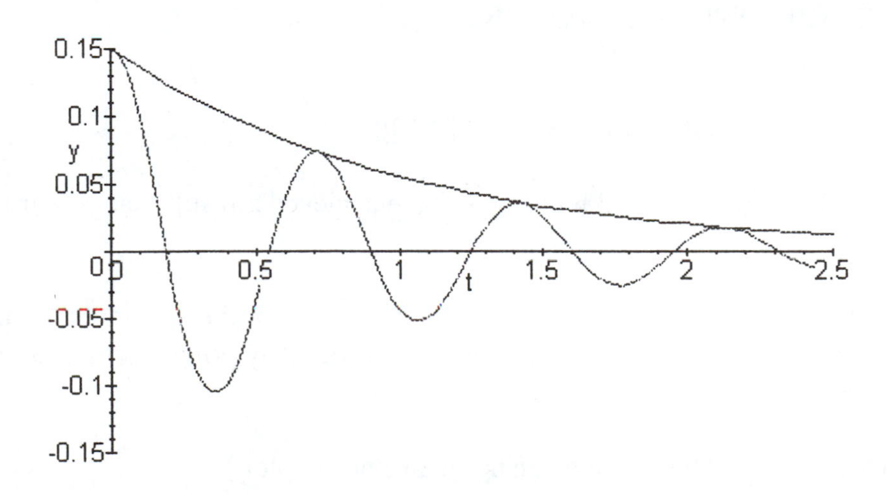

This observation will provide us with the necessary tool to make good estimates of A and b from the data set. The trick is to use our old friend the semi-log plot. If we make a semi-log plot of our data set, then the peaks of the oscillations should form what sort of geometric object?? Why? Place your answers below.

• Maple Tip. To produce a semi-log plot, use the command semiLogList to generate the data, and then plot it. The semiLogList command was created for this lab and acts as follows:

semiLogList([[x1,y1], [x2,y2], ...]) = [[x1,log(y1)], [x2,log(y2)], ...]

A problem with taking the log of the y-coordinates of the xTimeData is that roughly half those coordinates are negative and the log function produces real outputs only for positive numbers. On the other hand, our plot command will not plot complex values, so in effect we will be plotting only the parts of the graph that appear above the x-axis. Enter the following command.

```
> semiLogData := semiLogList (xTimeData):
```

Now create a semi-log plot of the data set xTimeData with positive y-coordinates with the plot command applied to semiLogData. Enter the following set of commands.

```
> semiLogPlot :=plot(semiLogData, color=red):";
```

Do you think this gives evidence of exponential decay of the xTimeData? Do the peaks appear as you expected in the semi-log plot? Place your answer below.

We will now determine the coefficients of the straight line which best approximates the peaks in the semi-log plot. These coefficients should, in turn, lead us to values for what other quantities? Place your answer below.

Roughly estimate, from the semi-log plot, values for the coefficients a and b for the best line a*t + b passing through the peak points. Place these values into the following graphics command. Complete and Enter the following commands.

```
> a := ???: b := ???:
> plots[display]([semiLogPlot, plot(a*t + b, t=0..10)]);
```

Continue to vary the values of a and b until you have achieved a result that you are satisfied with. Enter your commands above.

We wish to use the values of a and b to obtain values for two of the constants in the modeling function. To do this, consider the following bit of algebra. Any point (t,x) which falls on the line a*t + b must satisfy the equation

log(x) = a*t + b (This is the meaning of a semi-log plot.)

Hence, taking the exponential with base E of each side of this equation yields

x = E^(a*t + b) = E^b * E^(a*t)

This, however, must be the formula for the damping envelope function. This gives us formulas for two of the constants in the modeling function x(t). Enter these formulas as Maple statements. Enter your commands below.

```
> A := evalf(???);
> beta := ???;
```

¤ f. Noting x0 and the shift d.

The initial displacement x0 can be read off from the first data point in xData. The [ ] command proves handy for this purpose. Enter the following command.

```
> x0 :=xData[1];
```

Now define the initial displacement x0. Place your definition below.

```
> x0 := ???
```

x(t) = A*exp(-b*t)*cos(w*t + d),

for some constants A, b, w, and d.

The quantity d =Delta represents a (generally small) shift to the left or right along the time axis of the oscillations. This comes from the term cos(w*t + d) in our formula for x(t). This parameter is best handled by taking the initial value of d to be zero. We will adjust this below when considering the graph of our proposed modeling function. Enter the following command.

```
> Delta := 0;
```

¤ g. Graphing and refining the model function.

You should now have values defined for the five constants in the following list. Enter this list. Be sure that you get numerical values for each entry.

```
> [A, beta, w, Delta, x0];
```

In Section 2 we defined the Maple function x(t) to be the modeling function given in terms of the constants A, beta, w, and Delta. Make sure that x(t) remains defined in this way (either reEnter the definition for x(t) in Section 2 or copy it below and reEnter it). Complete and Enter the command.

```
> x:= t -> ???;
```

We are now ready for the big test! We have determined values for all the constants that appear in the modeling function x(t), and so we can plot it. In particular, we can plot x(t) along with a plot of our data set. Enter the following commands.

```
> A := A; b := beta;  w := 5.8; d := Delta;
> modelPlot := plot(x(t),t=0..10, y=-.23.. .23, color=blue):
> plots[display]([dataPlot, modelPlot],  title=`Model & Data Compared`);
> x(t);
> T := evalf(2*Pi/5.8);
```

How good a match did you obtain? You should now refine your values of the four constants by reEntering the above command as often as necessary. Stop when you are satisfied with the results. Enter your commands above.

¤ h. Estimating the damping constant and the spring constant.

Now that we have values for the constants in the modeling function x(t), we can go back one more step and determine the constants in the original differential equation,

d^2x/dt^2 + (Alpha/m)*(dx/dt) + (k/m)*x = 0,
where x(0) = x0, and dx/dt(0) = 0.

Recall that Alpha is the damping constant, k the spring constant, and x0 the initial displacement. We already have the value of one of these constants--which one? Place your answer below.

However, in Section 2 you will find equations which can be manipulated into formulas for the remaining two constants. Derive these formulas and Enter them as Maple commands. The results should be numerical values for all of the constants appearing in the differential equation which models the damped spring. Enter your commands below.

Note: The units of measurement for these constants will depend on the units chosen for length, mass, and time. In order to avoid unnecessary detail, we have not attempted to specify units except for mass (kg) and time (sec).

¤ i. The final comparison!

We have now come full circle, back to the original differential equation model for the damped spring. Since you now have values for all the constants in this differential equation, you can generate numerical solutions using the command dampedSpring. Do this below, picking n and Deltat so that the numerical approximation will cover the first 10 seconds of the spring's motion. Complete and Enter the following command.

```
> numPlot := dampedSpring(???):";
```

Does your numerically generated curve match the data plot? Let's check the match with the following (ultimate) plot. Enter the following command.

```
> plots[display]([numPlot, modelPlot, dataPlot],title=`x(t) = blue, Data = red, Euler = green`);
>
```

Do you think you have achieved a good understanding of the spring/mass system which generated our data set? Why? Place your answer below.

¤ 5. Another data set...and a question. (Optional)

Duke has supplied us with another data set for a damped spring/mass system. The same mass of 550kg was used in both runs, but we were not told if the same spring was used both times. However, we were told that no cardboard was attached to the mass during the second run, so that the damping effect of air resistance will be much less than for the first run. Enter the following data set.

```
> heightData2 := [ 0.606, 0.606, 0.609, 0.616, 0.626, 0.639, 0.655, 0.674, 0.694, 0.716, 0.738, 0.
> 761, 0.783, 0.804, 0.826, 0.844, 0.857, 0.869, 0.880, 0.886, 0.888, 0.886, 0.880, 0.872, 0.858, 0
> .845, 0.828, 0.809, 0.787, 0.765, 0.743, 0.719, 0.700, 0.679, 0.661, 0.643, 0.629, 0.619, 0.612,
> 0.608, 0.608, 0.612, 0.619, 0.630, 0.643, 0.659, 0.678, 0.700, 0.722, 0.743, 0.766, 0.788, 0.808,
>  0.827, 0.845, 0.859, 0.870, 0.880, 0.884, 0.886, 0.883, 0.877, 0.868, 0.854, 0.839, 0.822, 0.804
> , 0.782, 0.760, 0.738, 0.716, 0.695, 0.676, 0.657, 0.642, 0.630, 0.620, 0.613, 0.610, 0.611, 0.61
> 5, 0.622, 0.633, 0.647, 0.663, 0.681, 0.701, 0.723, 0.747, 0.769, 0.790, 0.810, 0.829, 0.845, 0.8
> 59, 0.870, 0.878, 0.883, 0.883, 0.881, 0.874, 0.866, 0.853, 0.838, 0.821, 0.801, 0.781, 0.759, 0.
> 738, 0.716, 0.696, 0.675, 0.657, 0.643, 0.630, 0.621, 0.615, 0.613, 0.613, 0.618, 0.626, 0.637, 0
> .651, 0.667, 0.685, 0.705, 0.726, 0.747, 0.769, 0.790, 0.812, 0.830, 0.843, 0.856, 0.869, 0.877,
> 0.881, 0.881, 0.877, 0.872, 0.862, 0.849, 0.835, 0.817, 0.798, 0.778, 0.757, 0.735, 0.714, 0.695,
>  0.675, 0.659, 0.644, 0.633, 0.624, 0.619, 0.617, 0.619, 0.624, 0.632, 0.643, 0.657, 0.673, 0.692
> , 0.711, 0.732, 0.753, 0.774, 0.794, 0.814, 0.831, 0.846, 0.859, 0.869, 0.874, 0.878, 0.878, 0.87
> 4, 0.867, 0.858, 0.843, 0.827, 0.810, 0.791, 0.771, 0.750, 0.730, 0.709, 0.690, 0.671, 0.656, 0.6
> 42, 0.631, 0.623, 0.618, 0.619, 0.621, 0.626, 0.635, 0.646, 0.661, 0.677, 0.696, 0.715, 0.736, 0.
> 756, 0.777, 0.797, 0.816, 0.832, 0.847, 0.859, 0.868, 0.874, 0.877, 0.876, 0.872, 0.865, 0.855, 0
> .842, 0.827, 0.809, 0.790, 0.770, 0.750, 0.729, 0.709, 0.688, 0.670, 0.655, 0.642, 0.632, 0.624,
> 0.620, 0.619, 0.621, 0.627, 0.635, 0.647, 0.662, 0.678, 0.696, 0.716, 0.736, 0.757, 0.779, 0.799,
>  0.817, 0.833, 0.847, 0.858, 0.867, 0.873, 0.874, 0.874, 0.869, 0.862, 0.851, 0.838, 0.823, 0.806
> , 0.786, 0.767, 0.746, 0.726, 0.707, 0.688, 0.671, 0.655, 0.642, 0.632, 0.625, 0.622, 0.621, 0.62
> 4, 0.630, 0.639, 0.651, 0.666, 0.682, 0.700, 0.719, 0.739, 0.759, 0.779, 0.798, 0.817, 0.833, 0.8
> 47, 0.858, 0.865, 0.870, 0.872, 0.871, 0.866, 0.858, 0.847, 0.834, 0.819, 0.801, 0.783, 0.761, 0.
> 741, 0.722, 0.702, 0.685, 0.668, 0.654, 0.642, 0.633, 0.627, 0.624]:
>
```

The question you must answer is this: can you prove whether or not the same spring was used in both runs? Be sure to supply enough computations to vigorously support your claim. Enter your computations and explanations below.

□ LABORATORY SUMMARY SHEET.

Modest damping (a >0).

k=4; m=.1; x0=.15;

Alpha	initial amplitude	amplitude decline	period
____	________	________	________
____	________	________	________
____	________	________	________
____	________	________	________
____	________	________	________
____	________	________	________

Conclusions.

How does a affect amplitude?

Can amplitude decline ratio be made arbitrarily small?

How does a affect the period?

Can the period be made arbitrarily large by adjusting a? □

LAB 13A: Projectile Motion

□ NAMES:

□ PURPOSE.
In this laboratory exercise we investigate Euler's Method approximations to the solution of the projectile problem (as described in Chapter 5 of Calculus: Modeling and Application), with special attention to air resistance and the effect of a head wind. We must use a numerical solution technique such as Euler's Method since most of the problems we consider in this lab are impossible to solve with purely symbolic calculations.

□ PREPARATION.
Read Chapter 5 Lab Reading in Calculus: Modeling and Application, pages 347 - 352, and answer the questions found in this material.

□ LABORATORY REPORT INSTRUCTIONS.

Note: Follow your instructor's instructions if they differ from those given here.

For this laboratory you are not required to submit a formal report. Rather you are asked to submit a copy of this Worksheet with the questions answered in the indicated locations. Only one submission will be made. Write in complete sentences and be sure the sentences connect together in coherent ways.

□ Maple Procedures.
Be sure to Enter the following command before beginning the lab.

```
> with(lab13a);
> # If an Error is an output of the above line, see your lab instructor.
```

Three new commands are defined for this lab: Projectile, ProjectilePlotter, and TimeFinder; these will be the main tools. These commands are described later in the Worksheet.

□ PRELIMINARIES.

¤ 1. Description of the Problem.

The situation is as described in Chapter 5 of Calculus: Modeling and Application. The Northerners are at the top of a hill 400 meters above the plain on which the city of Ergo is situated, and 600 meters (horizontal distance) away from the center of the city. To drop their leaflets over Ergo, the Northerners must burst a shell at a height of 100 meters directly over the center of the city.

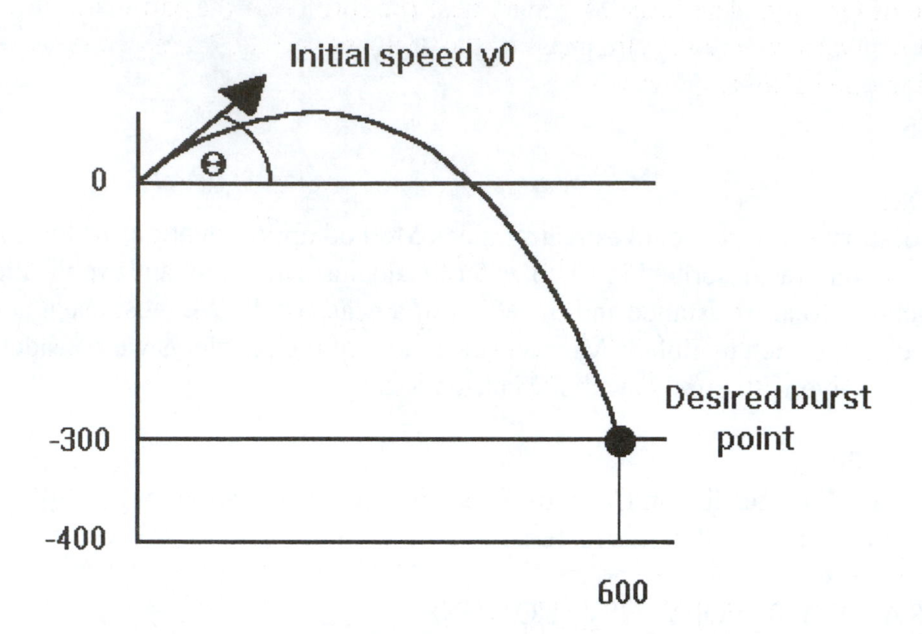

We designate the top of the hill as the origin of the x, y coordinate system, i.e., x=0 and y=0. Then the desired shell burst point has coordinates x=600 and y=-300. The shell is launched at a speed of v0=100 meters/sec and at an angle Theta with the horizontal. We further assume that air resistance on the shell is proportional to the square of the shell's speed.

The justification for assuming this form of the retarding force is given in Calculus: Modeling and Application.

The primary goal of this lab is to determine two quantities: the angle Theta that will send the shell through the desired burst point, and the time t at which the shell reaches this location. We will numerically approximate these values by applying Euler's Method to the initial value problems which model the motion of the shell.

We then consider how changes in the model (e.g., a head wind) will affect the values of Theta and t.

¤ 2. Derivation of the Initial Value Problem Model.

Let x = x(t) and y = y(t) denote the coordinates of the position of the shell at time t.

The symbols for the horizontal and vertical velocities are vx and vy (m/sec).

The symbols for the horizontal and vertical accelerations are ax and ay (m/sec^2).

There are two forces on the shell that we wish our model to take into account: the force of gravity and the drag force of air resistance.

The Force of Gravity. The force of gravity near the surface of the earth is well-known: mg, pointed downward, where m is the mass of the shell and g=9.807 m/sec^2 is the component of acceleration due solely to gravity.

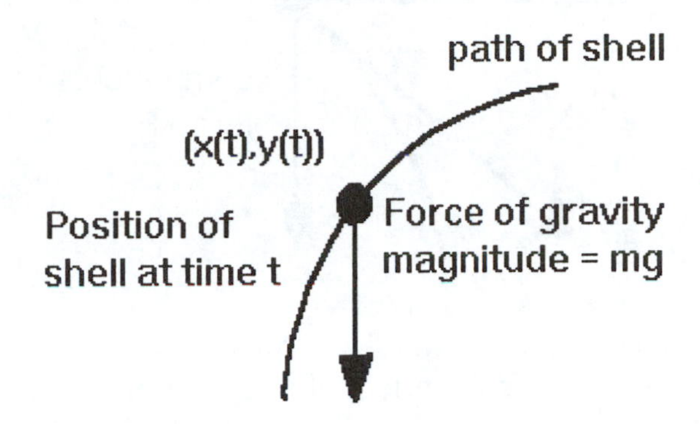

The Drag Force of Air Resistance. As described in the previous section, we assume that the drag force is proportional to the speed squared and directed opposite to the direction of motion. Hence, if |v| denotes the speed of the shell at time t, then there exists a constant c0 such that the magnitude of the drag force is c0*|v|^2. Here is a picture to make this clear.

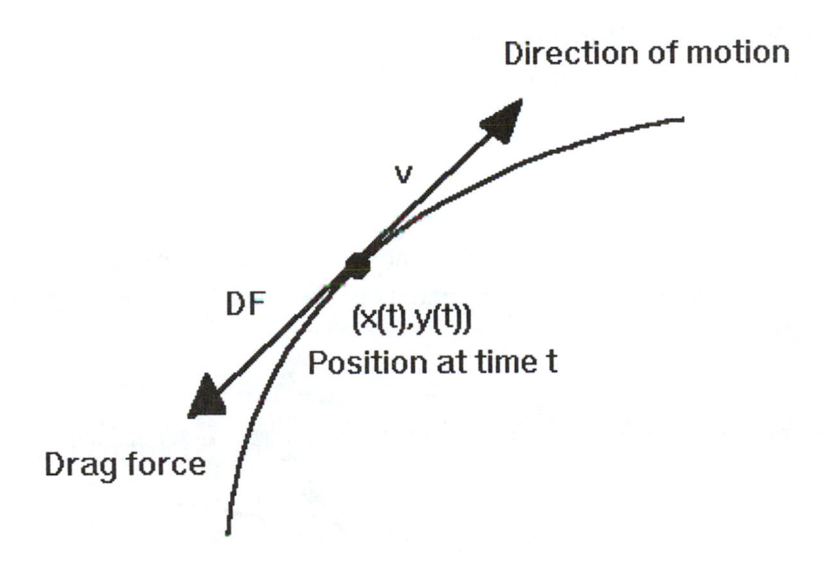

Components of a Force. In order to analyze the motion of the shell we need to determine the effects of the two forces (gravity and air resistance) together on the shell. The problem is that they act in different directions. How do we handle this difficulty?

The solution is to break both of the forces down into horizontal and vertical components. Any force F can be broken down into horizontal and vertical components Fx and Fy as shown below.

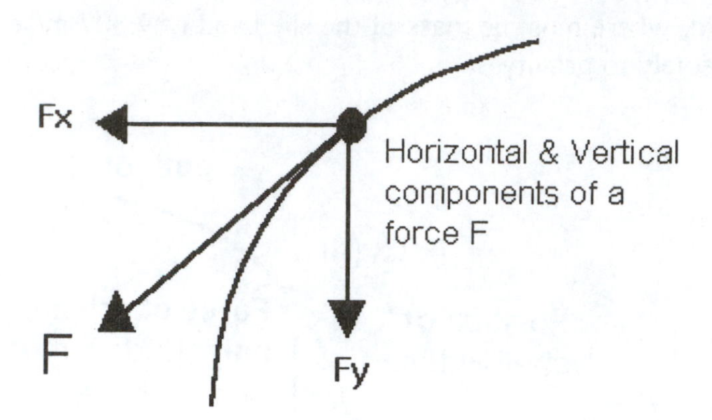

It is an observed law of physics that the effect of a force F on an object is the combination of the effect of the horizontal component force Fx and the effect of the vertical component force Fy. Hence we now determine the components of both the force of gravity and the force of air resistance.

Components of the Force of Gravity. Gravity has only a vertical component. Complete the following box on paper, then activate the two lines which follow.

Horizontal component of force of gravity: ??
Vertical component of force of gravity: ??

```
> Result(1);
>
```

Components of the Drag Force of Air Resistance. These are harder to determine. Consider the following picture, showing both the components of the drag force vector and the components of the velocity vector.

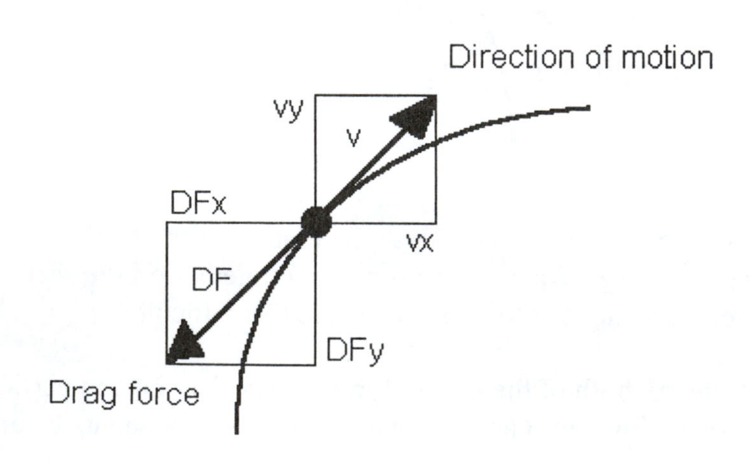

Notice that the "force rectangle" is merely an enlarged version of the "velocity rectangle." This means that there is some number ct such that every piece of the "force rectangle" is ct times the corresponding piece of the "velocity rectangle." Applying this to the three pairs of arrows shown in the picture above gives... Complete the following equations on paper, then activate the line which follows.

c0*|v|^2 = ct*|v|
Fx = ???
Fy = ???

```
> Result(2);
```

However, the first of these equations gives us a formula for ct in terms of c0 and |v|: Complete the following equation on paper, then activate the line which follows.

ct = ???

```
> Result(3);
```

Hence, placing this formula for ct into the formulas for Fx and Fy result in: Complete the following equations on paper, then activate the line which follows.

Fx = ???
Fy = ???

```
> Result(4);
```

So, taking into account the opposite direction of the drag force to the velocity, we have the desired components of the drag force of air resistance: Complete the following box on paper, then activate the line which follows.

Horizontal component of drag force: ???
Vertical component of drag force: ???

```
> Result(5);
```

Components of the Total Force. The components of the total force on the shell are just the additions of the corresponding components for the force of gravity and the drag force: Complete the following box on paper, then activate the line which follows.

Horizontal component of total force: ???
Vertical component of total force: ???

```
> Result(6);
```

Obtaining the Differential Equation Models. Newton's Second Law of Motion states that force equals mass times acceleration. The trick here is to apply Newton's Law twice, once in the horizontal direction and once in the vertical direction. The forces in these two directions are merely the horizontal and vertical components of the total force as obtained in the previous paragraph. Complete the following box on paper, then activate the line which follows.

mass times acceleration equals force
Horizontal component: m*ax = ???
Vertical component: m*ay = ???

```
> Result(7);
```

Divide each equation by m, and replace ax and ay with dvx /dt and dvy /dt. This yields... Complete the following box on paper, then activate the line which follows.

Horizontal component: ???
Vertical component: ???

```
> Result(8);
```

Let c = c0/m and note that |v| = sqrt(vx^2 + vy^2) by the Pythagorean Theorem.

Then our two differential equations can be rewritten in the following form... Complete the following box on paper, then activate the line which follows.

Horizontal component: dvx /dt = ???
Vertical component dvy /dt = ???

```
> Result(9);
```

The Initial Conditions. It was assumed that the initial speed of the shell was 100 m/sec, and the initial direction of the shell was set at an angle of Theta to the horizontal:

Hence, using the definitions of the sine and cosine, we obtain the initial values for vx and vy... Complete the following equations on paper, then activate the line which follows.

vx(0) = ??? and vy(0) = ???

```
> Result(10);
```

Combining these with our differential equations for vx and vy give us the initial value problem model for the motion of the shell... Complete the following lines on paper, then activate the command which follows.

Horizontal: dvx /dt = ??? with vx(0) = ???
Vertical: dvy /dt = ??? with vy(0) = ???

```
> Result(11);
```

These are the equations on the bottom of page 351 of Calculus: Modeling and Application.

¤ 3. Deriving the "Updating Equations" for Euler's Method.

Given a fixed value of Theta, we will now use Euler's Method to approximate the two velocity functions, vx=vx(t) and vy=vy(t), and the two position functions, x=x(t) and y=y(t). Starting with a step size of Dt=0.1 seconds, we will apply Euler's Method a sufficient number of times to approximate the full trajectory of the shell. The values of the quantities so produced at time t = k*Dt will be denoted by vx(k), vy(k), x(k), and y(k).

Euler's Method--a Review. Suppose we have a time dependent quantity s=s(t) that we wish to approximate by Euler's Method using a time step size of dt. (This is what we will do with vx, vy, x, and y.) The "updating equation" for Euler's Method is given by:

s(k+1) = s(k) + (slope at s(k))*dt

This formula is explained by the following diagram:

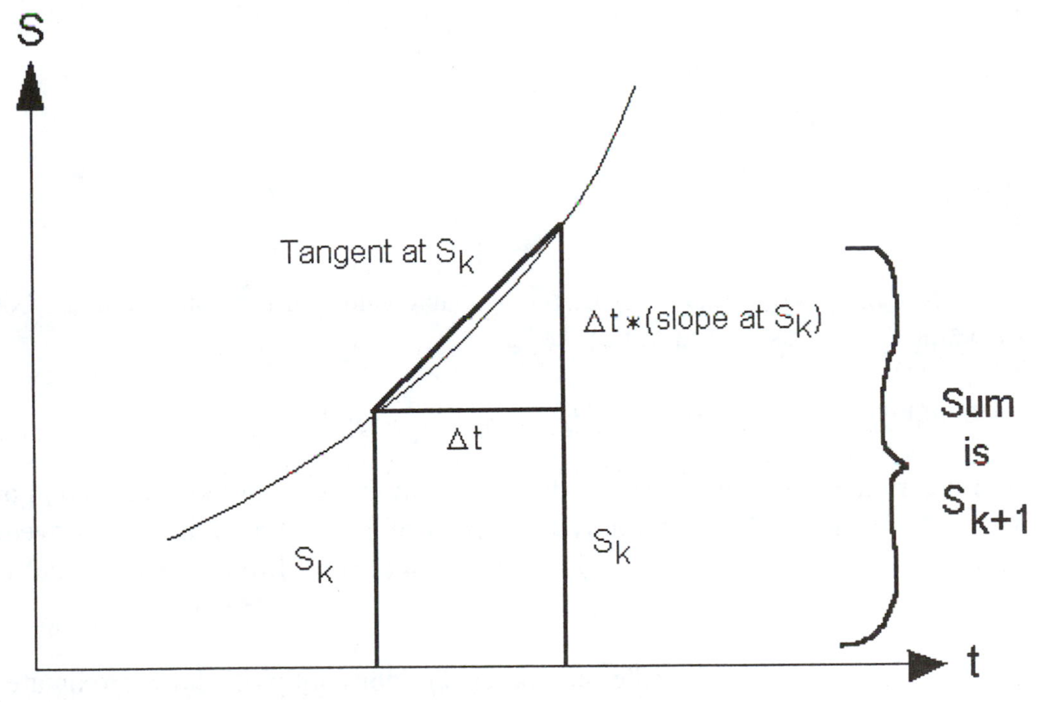

The slope at s(k) is merely the slope of the tangent at s(k). When this slope is multiplied by the run (which is dt), it gives us the rise, the term we must add onto s(k) to obtain s(k+1). Since the slope of the tangent at s(k) is merely ds/dt at this point, the "updating equation" for Euler's Method becomes:

s(k+1) = s(k) + (ds/dt(sk))*dt

Apply this generic updating equation to the quantities vx, vy, x, and y and obtain equations that show how to go from

vx(k), vy(k), x(k), and y(k)
to
vx(k+1), vy(k+1), x(k+1), and y(k+1).

Hint: We have obtained formulas for the derivatives of vx and vy, and we know that the derivatives of x and y are vx and vy respectively. Use dt to stand for Dt. Complete the following equations on paper, then activate the line which follows.

```
vx(k+1) = vx(k) + ( ???                    )*dt
vy(k+1) = vy(k) + ( ???                    )*dt
 x(k+1)  =  x(k) + ( ???                   )*dt
 y(k+1)  =  y(k) + ( ???                   )*dt
```

> Result(12);

For a fixed value of the angle Theta and initial speed v0, write down the initial values for the four quantities vx, vy, x, and y. Complete the following equations on paper, then activate the line which follows.

```
vx(0) = ??        and  vy(0) = ??
 x(0) = ??        and   y(0) = ??
```

> Result(13);
>

For shells of the form used in this problem, a reasonable value for the constant c is 0.000072. This is the value we will use in the rest of the lab.

¤ 4. Implementing Euler's Method.

We wish to determine the angle Theta and the time t that the Northerners will need in order to burst their shell at exactly 100 meters above the center of Ergo. To do this it is convenient to construct a few new Maple functions. This is done with the Lab 13A Procedures that were activated at the beginning of this lab.

The three new commands are Projectile--computes (but does not plot) the approximate values vx(k), vy(k), x(k), and y(k) using the updating equations; ProjectilePlotter--plots the x,y trajectory, along with a t,y graph; and TimeFinder--prints a table of y and t values near a specified t value. More information on these functions can be found by using the ? command. They will be described more thoroughly as needed in the Project.

□ PROJECT.

¤ 1. Determining Theta and t in the absence of a headwind.

There is no headwind when the Northerners arrive at the top of the hill. Thus all the equations derived in the Preliminaries section can be used without alteration--these are built into the commands Projectile and ProjectilePlotter. We will use these commands to determine the angle Theta that causes the shell to pass through the desired target point.

The following statements are provided for your convenience. They plot the trajectory of the shell for your choice of the angle Theta (an Euler's Method approximation). They also plot the height of the shell as a function of time. (This will help us later determine the exact time that the shell reaches the desired burst point.) Use these statements to determine the angle Theta which sends the shell through the burst point. Enter your computations below.

```
> Theta := ???*degrees:
> projectileData1 := Projectile(Theta):
> ProjectilePlotter(projectileData1);
>
```

Given that you have found the angle Theta that sends the shell through the burst point, then use the plot of y as a function of time to estimate tEst, the time it takes to reach the burst point. Enter this value for tEst below, along with the TimeFinder command. The result will be a table of the y values for t near tEst (taken from the Euler's Method data generated to plot y versus t). Examine the values in this table to refine your value of tEst. Enter your computations below.

```
> tEst := ???:
> TimeFinder(tEst, projectileData1);
```

Record your results in the following spaces.

Given the drag force of air resistance, the angle of launch should be
??? degrees,

and the time required to reach the desired burst point will be
??? seconds.

Briefly describe your procedure for using TimeFinder to determine the time required to reach the desired burst point. Place your answer below.

Answer:

¤ 2. Determining Theta and t in the presence of a headwind.

Just as the Northerners are about to launch, they realize that a headwind has come up, blowing in from Ergo with a velocity of w = 10 meters per second. However, they came prepared for such an inconvenience: they brought a Notebook PC, a copy of Maple, the worksheet in which they did their original calculations, and a wicked big battery. They must modify the IVPs (and the corresponding updating equations for Euler's Method) so that the headwind w can be properly accounted for.

Fortunately the Northerners had done their homework and had answered Exploration Activity 2 at the end of Chapter 5 in Calculus: Modeling and Application -- this gave them the desired equations. Now you obtain these equations. For your convenience we include below the equations obtained when no headwind is present. (Hint: If the shell has x-velocity vx and y-velocity vy, then the headwind of w will make the shell feel like the x- and y-velocities are now equal to __________ and __________.) Make your modifications directly in these lines.

```
d(vx)/dt = - c*sqrt(vx^2+vy^2)*vx,  where  vx(0) = v0*cos(Theta)
d(vy)/dt = - g - c*sqrt(vx^2+vy^2)*vy,  where  vy(0) = v0*sin(Theta)

vx(k+1) = vx(k) + (  -c*sqrt(vx(k)^2+vy(k)^2)*vx(k))*dt
vy(k+1) = vy(k) + (-g -c*sqrt(vx(k)^2+vy(k)^2)*vy(k))*dt
x(k+1)  =  x(k) + vx(k)*dt
y(k+1)  =  y(k) + vy(k)*dt
```

Your modifications to the equations should be of the following form, where w1, w2, w3, w4, w5, and w6 are simple expressions in hw (including, possibly, zero):

```
d(vx)/dt = - c*sqrt((vx+w1)^2+(vy+w2)^2)*(vx+w3),
    vx(0)=v0*cos(Theta)
d(vy)/dt = - g - c*sqrt((vx+w4)^2+(vy+w5)^2)*(vy+w6),
    vy(0)=v0*sin(Theta)

vx(k+1) = vx(k) +(-c*sqrt((vx(k)+w1)^2+(vy(k)+w2)^2)*(vx(k)+w3))*dt
vy(k+1) = vy(k)+(-g-c*sqrt((vx(k)+w4)^2+(vy(k)+w5)^2)*(vy(k)+w6))*dt
x(k+1)  =  x(k) + vx(k)*dt
y(k+1)  =  y(k) + vy(k)*dt
```

Given any headwind hw, what were the choices you made for w1, ..., w6? Record these below. Complete and Enter the following equations.

```
> w1 := ???;   w2 := ???;   w3 := ???;
> w4 := ???;   w5 := ???;   w6 := ???;
```

The command Projectile has been written so as to allow for consideration of a headwind. A headwind hw can be added into the command via the options shown in the following line.

projectileData := Projectile(Theta, WindSpeed=10, WindSpeedModel=[values for w1,...,w6]):

As you see, you add in your values for w1, ..., w6. (So if you have the wrong relationships, you'll get some pretty strange plots....) As done in Section 1, use Projectile and ProjectilePlotter to determine the angle Theta that will cause the shell to pass through the desired target point--but now assuming a headwind of 10 m/sec. Be sure your formulas for w1, ..., w6 have been Entered above. Place your computations below.

```
> Theta := ???*degrees;  hw := 10;
> projectileData2 := Projectile(Theta, WindSpeed=hw, WindSpeedModel=[w1,w2,w3,w4,w5,w
> 6]):
> ProjectilePlotter(projectileData2);
>
```

Given that you have found the angle Theta that sends the shell through the burst point in the presence of a headwind of 10 m/sec, then estimate tEst, the time it takes to reach the burst point. Enter this value for tEst below, along with the TimeFinder command. Use the results to refine your value of tEst. Enter your computations below.

```
> tEst := ???:  TimeFinder(tEst, projectileData2);
>
```

Record your results in the space below.

In the presence of a 10 m/sec headwind, the angle of launch should be
??? degrees,
and the time required to reach the desired burst point should be
??? seconds.

How much difference does the wind make in your answers? If the Northerners do not take the headwind into consideration, will their mission succeed or fail? Why? (Hint: Determine where the shell will burst if, in the presence of a headwind, the Northerners decide to use the Theta and t values calculated without a headwind.) Enter your answers and computations below.

```
> Theta := ???*degrees;  hw := 10;
> projectileData3 := Projectile(Theta, WindSpeed=hw,WindSpeedModel=[w1,w2,w3,w4,w5,w6
> ]):
> ProjectilePlotter(projectileData3);
> TimeFinder(???,projectileData3);
>
```

Answer:

The Northerners have run into a bit of bad luck: just as they get to the top of the hill, Hurricane Legins hits, packing headwinds of 75 meters per second. Now what is the angle Theta that will cause the shell to pass through the desired target point? And at what time t will the target be hit? Enter your answers and computations below.

```
> Theta := 'Theta':
> Theta := ???*degrees ; hw := 75;
> projectileData4 := Projectile(Theta, WindSpeed=hw,WindSpeedModel=[w1,w2,w3,w4,w5,w6
> ]):
> ProjectilePlotter(projectileData4);
> Theta;
> tEst := ???; TimeFinder(tEst, projectileData4);
```

Record your results in the spaces below.

In the presence of a 75 m/sec headwind, the angle of launch should be
??? degrees,

and the time required to reach the desired burst point should be
??? seconds.

How much difference does the wind make in your answers? In particular, if the Northerners do not take the hurricane headwind into consideration, will their mission succeed or fail? Why? (Hint: Determine where the shell will burst if, in the presence of a hurricane headwind, the Northerners decide to use the Theta and t values calculated without a headwind.) Enter your computations and answers below.

```
> Theta := ???*degrees ; hw := 75;
> projectileData5 := Projectile(Theta, WindSpeed=hw, WindSpeedModel=[w1,w2,w3,w4,w5,w
> 6]):
> ProjectilePlotter(projectileData5);
>
> tEst := ???; TimeFinder(tEst, projectileData5);
```

Answer:

¤ 3. Can the Northerners ignore the drag of air resistance?

Ignoring the drag of air resistance in our model is equivalent to setting the "velocity squared" constant c equal to 0. Explain why this is so. Place your answer below.
Answer:

The command Projectile has been constructed to allow for the model with no air resistance by adding in the option shown in the following line.

```
projectileData := Projectile(Theta, VelocitySquaredConstant=0);
```

As done in Sections 1 and 2, determine the angle Theta that will cause the shell to pass through the desired target point, and the time t at which this will occur--but now assuming no air resistance. Also assume no headwind is present. Enter your computations below.

```
> Theta := ???*degrees ;
> projectileData6 := Projectile(Theta, VelocitySquaredConstant=0):
> ProjectilePlotter(projectileData6);
> tEst := ???; TimeFinder(tEst, projectileData6);
>
```

Record your results in the spaces below.

With no drag from air resistance, the angle of launch should be
??? degrees,
and the time required to reach the desired burst point should be
??? seconds.

How does the answer obtained here compare with the answer obtained in Exploration Activity 2 at the end of chapter 5 in Calculus: Modeling and Application? Does this give you confidence in your methods? Place your answer below.
Answer:

Conjecture: "A headwind, coming in at a horizontal direction, will only affect the velocity in the x direction, and not the velocity in the y direction." Referring to the equations derived in this section, either support or refute this conjecture. Place your answer below.
Answer:

In this model--where we assume no air resistance--are assumptions about the presence or absence of a headwind important, i.e., would the presence of a headwind change the answers you generated above? The justification for your answer should include references to the nature of the equations we have used to model the motion of the shell. Place your answer below.
Answer:

How much difference does the drag force make in your answers? In particular, if the Northerners do not take the drag force into consideration, will their mission succeed or fail? Why? (Hint: Determine where the shell will burst--according to the air resistance model--if the Northerners decide to use the Theta and t values calculated without considering air resistance. Assume no headwind is present.) Enter your computations and answers below.

```
> Theta := ???*degrees ;
> projectileData7 := Projectile(Theta):
> ProjectilePlotter(projectileData7);
> tEst := ???; TimeFinder(tEst, projectileData7);
>
```

Answer:
□

LAB 13B: Rocket Motion

□ NAMES:

□ PURPOSE

We continue to explore the use of Euler's Method to solve differential equations, this time the rocket motion equations developed in Calculus: Modeling and Application. We apply the solution to explore the advantages of a two-stage rocket over a one-stage rocket. In this problem, in contrast to the projectile problem considered in a previous lab, both the gravity and the mass vary with time.

□ PREPARATION

Read project 1 at the end of Chapter 5 in Calculus: Modeling and Application, and at least work parts k and l.

□ LABORATORY REPORT INSTRUCTIONS

Note: Follow your instructor's instructions if they differ from those given here.

For this project you are not required to submit a formal report. Rather you are asked to submit a copy of this Worksheet with the questions answered in the indicated locations. You will be graded on this single submission. Write in complete sentences and be sure the sentences connect together in coherent ways.

□ Maple Procedures.

```
> with(lab13b);
> # If an Error is an output of the line above, see your lab instructor.
```

These procedures define two new commands: RocketMotion and VelocityPlot. These will be described in more detail as needed in the lab.

□ PRELIMINARIES.

¤ 1. Determining Escape Velocity.

The situation is as described in Calculus: Modeling and Application, Chapter 5 Project 1. Starting at a distance d from the center of the earth, suppose a projectile is moving directly away from the center--"straight up." What initial velocity must the projectile have so that it will not return?

We are assuming that the projectile is "free-floating," i.e., that it has no propulsion of its own. (In this lab our "free-floating" projectile will be a rocket after its fuel has been totally consumed.) If we further ignore air resistance, then the earth's gravity is the only force on the projectile.

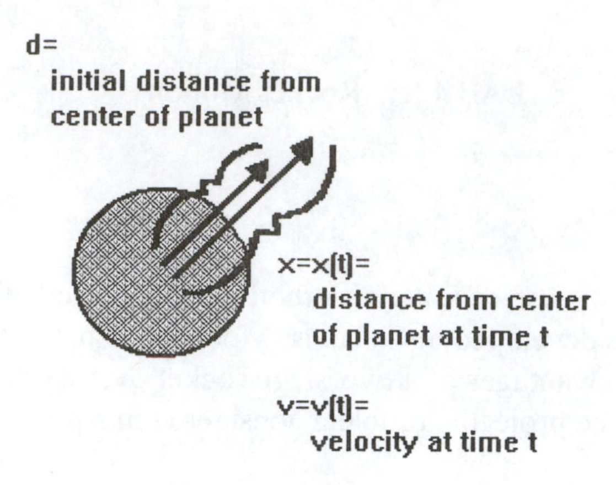

Under the above assumptions, Newton's Law of Gravitation and Newton's Second Law of Motion are used in Calculus: Modeling and Application to show

m*dv/dt = - G*m/x^2

where m is the mass of the projectile,

x=x(t) is its distance from the center of the earth at time t,

v=v(t) is its velocity at time t, and

G is a proportionality constant which does not depend on any of the other quantities. It is known that the value of G is 95,040 miles^3/seconds^2.

We will need one other variable:

d is the distance to the center of the earth when "free floating" flight begins.

Thus d=x(0). In particular, if the projectile started its "free floating" flight at the surface of the earth, then d equals 3960 miles.

Calculus: Modeling and Application shows that if the projectile started its "free floating" flight with an initial velocity of v0, then the velocity of the projectile as a function of the distance x during the rising portion of the trip is given by

v^2 = 2*G*(1/x - 1/d) + v0^2 .

Our Goal: Find the (minimum) escape velocity V0 as a function of the distance d, i.e., the smallest value of v0 such that v^2 never equals zero for any value of x greater than d. This will insure that the projectile escapes from the Earth's gravitational field.

Since v^2 is a strictly (decreasing, increasing) (choose one) function of x, then v^2 will be unequal to 0 for all values of x so long as it remains unequal to 0 for arbitrarily large values of x. This will be the case if the limit of v^2 as x becomes arbitrarily large is greater than or equal to 0, i.e., lim(x->inf) v^2 >= 0.
Why can we allow the value zero in the limit when we don't want it for any (finite) value of x? Decide on an answer, then activate the line which follows to check your claim.

```
> Answer(1);
```

Thus the minimum value of v0 which will insure that our projectile escapes from the Earth (i.e., the escape velocity V0) will be that value of v0 which makes lim(x->inf) v^2 equal to ???. Complete the following equation, then activate the next line.

lim(x->inf) v^2 = ???

```
> Answer(2);
```

Combining this limit equality with our formula for v^2 allows us to solve for the (minimum) escape velocity v0=V0. Write your answer on paper, then activate the line which follows.

V0 = ???

```
> Answer(3);
```

As you move further from the center of the earth before starting the "free floating" motion (i.e., increase d), what happens to the corresponding value of the escape velocity? How might this be relevant to rocket design when trying to send a payload to another planet? Decide on an answer, then activate the line which follows to check your claims.

```
> Answer(4);
```

¤ 2. Rocket Motion: the Starting Constants.

In this section we consider a rocket which is being propelled "upward" (i.e., away from the center of the earth) by the "push" of exhaust gases from burning fuel. Our goal is to determine if, at the end of the burn period (i.e., when all the fuel is consumed), the rocket will be able to escape from the gravitation field of the Earth. This is like the situation considered in Chapter 5 Project 2 of Calculus: Modeling and Application.

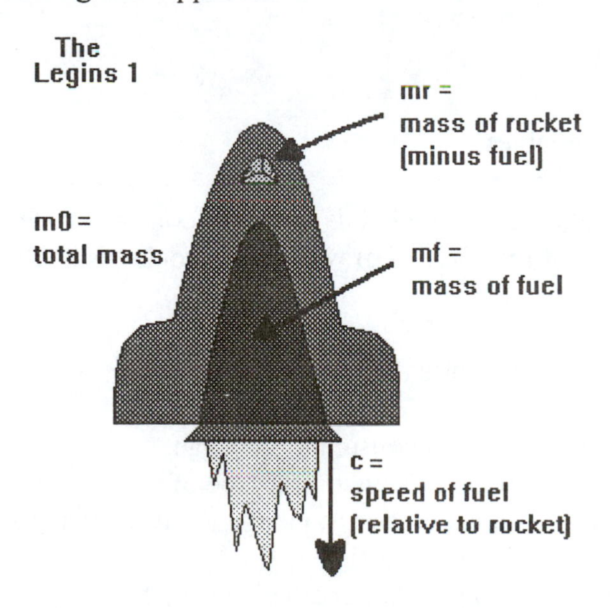

Here are the constants which enter our model:

m0 = the initial total mass of rocket and fuel.
mf = the initial mass of the fuel.
mr = the mass of the rocket.
c = the speed of fuel exhaust (relative to the rocket itself).
tBurn = the total length of burn, i.e., the time necessary for all the fuel to be consumed. We assume that the fuel is burned at a uniform rate over the time interval from t=0 to t=tBurn.
r = rate of fuel consumption (kg/sec) during the burn period.

Let's consider these six constants in more depth--we claim that we can trim the list down to just two unknown constants.

The mass constants m0 and mf, and mr are clearly related. Express mr in terms of m0 and mf. Write your answer on paper, then activate the line which follows.

```
> Answer(5);
```

We will later specify exact values for the constants c and tBurn, so these can be assumed to be known. The final constant r can be expressed in terms of mf and tBurn--do so. Write your answer on paper, then activate the line which follows.

```
> Answer(6);
```

What are the two remaining constants whose values we will need to specify? Write your answer on paper, then activate the line which follows.

```
> Answer(7);
```

¤ 3. Rocket Motion:

• The Time-dependent Variables.

We now consider the motion of the rocket as its fuel is consumed. For simplicity, during this analysis we will assume that the force of gravity on the rocket is negligible when compared with the force exerted by the fuel exhaust.

Here are the time dependent variables that we will need:

t = the time in seconds since fuel consumption began.
s = s(t), the height of the rocket above the surface of the earth at time t.
(Hence x = s + 3960 miles, where x is the distance to the center of the Earth.)
v = v(t), the velocity of the rocket at time t.
m = m(t), the total mass of the rocket and fuel at time t.

At time t=0, we have initial values s0, v0, and m0. At the end of the fuel consumption period (i.e., t = tBurn), we have terminal values st, vt, and mt. The issue is simply this: does vt equal or exceed the escape velocity associated with the height st? If it does, then what will happen to the rocket? Check your answer by activating the following line.

```
> Answer(8);
```

If, on the other hand, the terminal velocity vt does not exceed the escape velocity associated with the height st, then what will happen to the rocket? Check your answer by activating the following line.

```
> Answer(9);
>
```

Determining whether or not our rocket escapes from the Earth is therefore dependent on determining st and vt, the height and velocity of the rocket at the end of its fuel consumption. (The escape velocity will then be computed from st by the formula obtained in Section 1 of these Preliminaries.) Determining st and vt will be done via Euler's Method.

The key ingredient in our Euler's Method solution is the formula for dv/dt derived in Project 1 Chapter 5 of Calculus: Modeling and Application (through step j) via Conservation of Momentum (this was a non-trivial derivation!):

dv/dt = -(c/m)*dm/dt

We will now express the right side of this equation in terms of the constants from the previous section. The initial total mass is m(0) = m0; the only change in this quantity comes about from the consumption of fuel, which is assumed to occur at a constant rate during a time interval of length tBurn. Give a simple formula for dm/dt which is valid during the burn period. Write your answer on paper, then activate the line which follows.

```
> Answer(10);
```

Use your formula for dm/dt to give a formula for m = m(t) during the burn period. Write your answer on paper, then activate the line which follows.

```
> Answer(11);
```

The formulas for m and dm/dt now give us a useful formula for dv/dt. What is it? Write your answer on paper, then activate the line which follows.

```
> Answer(12);
```

This differential equation can be solved symbolically by finding an antiderivative for the function of t given on the right hand side of the equation. However, we choose to generate solutions numerically by the use of Euler's Method, as described in the next section.

¤ 4. Deriving the "Updating Equations" for Euler's Method.

We use Euler's Method to approximate the velocity function, v=v(t), as well as the position function, s=s(t). Starting with a step size of deltat seconds, we apply the Euler Method n times; the values of the quantities so produced at time t = t(k) = k*Deltat will be denoted by v(k) and s(k), where k goes from 0 to n.

The "updating equations" for Euler's Method are given by:

v(k+1) = v(k) + Deltat*(slope of v at v(k))
s(k+1) = s(k) + Deltat*(slope of s at s(k))

However, from the final result in the previous section, we can obtain a formula for the "slope of v=v(t) at vk." What is this formula? Write your answer on paper, then activate the line which follows.

"slope of v(t) at vk" = ????

```
> Answer(13);
```

Hence we can obtain the "updating equation" for v(k+1). What is this? (For convenience later, use k*Deltat in place of t(k) in your formula for v(k+1).) Write your answer on paper, then activate the line which follows.

v(k+1) = v(k) + ???

```
> Answer(14);
```

Now we consider the position function s = s(t). What is the physical meaning of the derivative of s = s(t)? In view of this fact, there is a trivial formula for the "slope of s(t) at sk." What is it? Place your answer below.

"slope of s(t) at sk" = ????

```
> Answer(15);
```

Hence we can obtain the "updating equation" for s(k+1). What is this? Write your answer on paper, then activate the line which follows.

s(k+1) = s(k) + ???

```
> Answer(16);
```

We now have updating equations for s(k+1) and v(k+1). To see them together, activate the following line.

```
> Answer(17);
```

Invoking these the appropriate number of times will take us from s0, v0 to st, vt (the values of s(t) and v(t) at the end of the fuel consumption period). Then we will be able to compare vt with the escape velocity (as in Section 1) associated with the height st. Hence, for future reference, record the escape velocity vEsc(k+1) which is associated with the height s(k+1). (Remember: x = s + 3960). Write your answer on paper, then activate the line which follows.

vEsc(k+1) = ???

```
> Answer(18);
```

□ PROJECT.

¤ 1. A One Stage Rocket.

• Values of the Constants.

We recall our basic rocket from the Preliminaries, and specify values for the six constants as discussed in Section 2 of the Preliminaries. Here are values for three of those constants.

(1) The total burn time tBurn will be 30 seconds,
(2) The fuel exhaust speed will be c = 2.683 miles/sec.
(3) The total weight of the rocket and fuel will start at m0 = 15000 kg.

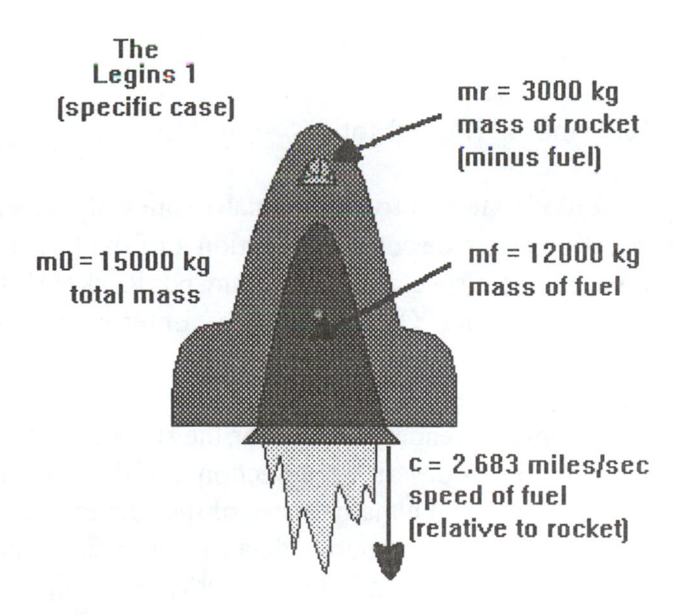

The remaining three constants from Section 2 are mf, mr, and r. However, recalling the discussion in Section 2, we know that we have only to specify the value of mf and the remaining constants will be determined. We now make a further assumption about the requirements of our rocket design that will lead to a value for mf.

In order to be able to carry a large amount of fuel and to burn it quickly, mr (the mass of the rocket minus fuel) must be roughly related to mf, the mass of the fuel. For the purposes of this lab we will assume that in order for the rocket to consume a fuel mass mf in a time span tBurn = 30 seconds, the rocket mass mr must be at least one-quarter of the fuel mass mf. Hence, taking mr to be as small as possible, we have

mr = (mf)/4

Thus, under all our assumptions, what are the exact values of all six constants? (Look back in Section 2 of the Preliminaries for the necessary equations.) For consistency use the units kg (kilograms), miles, and seconds (but do not list these units in the Maple definitions below). Complete and Enter the following statements.

```
> m0 := ???;   mr := ???;   mf := ???;
> c := ???;  tBurn := ???;   r := ???;
```

• The Time-dependent Variables.

We further assume that our rocket begins its flight at rest and on the surface of the Earth. This determines two of the initial values of the three time-dependent variables. The third initial value has already been specified. Enter your values for the three initial values below.

```
> s(0) := ???;
> v(0) := ???;
> m(0) := ???;
```

• Euler's Method--the Command "RocketMotion"

We are now ready to use Euler's Method to approximate values of s(t) and v(t) for 0 <= t <= tBurn. The necessary equations were discussed in Section 4 of the Preliminaries, and for your convenience are incorporated into a special Maple command: RocketMotion. Get further information on this command by using ?CommandName. Enter your information request below.

As you can see, RocketMotion only requires values for the two constants m0 and mf. The remaining constants are computed as described in Section 2 of the Preliminaries. Moreover, the default values for s(0), v(0), tBurn, and c match those of the rocket under study. Thus RocketMotion is currently configured to generate data for the rocket which we are considering. We have only to specify values of m0 and mf, and RocketMotion will do the rest.

As an example, execute RocketMotion for m0 = 15000 and mf = 12000. Complete and Enter the commands below.

```
> m0 := ???;   mf := ???;
> data := RocketMotion(m0, mf):
```

We can now construct a plot of the velocity and escape velocity data that has been produced by RocketMotion. For convenience--especially when later considering a two-stage rocket--we have a command VelocityPlot which produces the desired picture. (You can obtain further information on this command by using ?CommandName). Apply the command to the data list generated above by RocketMotion. Enter the following command.

```
> ?VelocityPlot
> VelocityPlot(data);
```

On the basis of your plot, does the rocket escape from the Earth's gravity? Explain. Place your answer below.

Answer:

RocketMotion has a number of options which you can alter--to see all of the options and their default values, execute the command Options(RocketMotion). In particular, you can alter BurnTime (tBurn) and FuelExhaustSpeed (c). (Whether we have the technological capability to actually alter these numbers for a real rocket is another issue). Can these quantities be altered sufficiently to change the answer to the previous question? Explain. Enter your commands and answers below.

• Maple Tip: Option values can be set inside a RocketMotion command in the way that we set option values in any Maple command, i.e., by OptionName=Value. As a specific example, if you wish to set c to 3.0 miles/second instead of the default 2.683 miles/second, your RocketMotion command would end with ..., mf, FuelExhaustSpeed = 3.0);

¤ 2. Designing a Two Stage Rocket.

We will now design a two-stage rocket such that the burned out second stage will carry a mp=50kg payload out of the earth's gravity. Here are the constraints that we must deal with:

• The total starting weight of the rocket shell, the fuel, and the payload, must be 15,000kg.

• For each of the two rocket stages, the weight of the stage (minus fuel) must be one-quarter of the weight of the fuel in that stage.

• Each stage of the rocket will burn its fuel at a uniform rate over a time interval of length tBurn=30 seconds.

• The exhaust gas from each stage will exit at a rate of c=2.683 miles/sec.

The first two constraints supply us with three linear equations in the five unknown constants mp, mr1, mr2, mf1, and mf2. Write down these equations, and then use them to express mr1, mr2, and mf2 in terms of mp and mf1. Write your answers on paper, then activate the line which follows.

```
> Answer(19);
```

Since we are given the value of mp, the only quantity left for us to adjust is mf1, the weight of the fuel in the first stage rocket. We will attempt to find values of mf1 which will result in our payload being taken out of the earth's gravity.

¤ 3. The Two Stage Rocket: a Specific Case.

Let's consider the case where mf1=7,000kg. Determine the values of the other quantities mp, mr2, mf2, and mr1. Enter all five of these quantities as Maple definitions. (You have explicit values for two of the quantities, and formulas for the remaining three.) Enter your commands below.

```
> mp := ??? :   mf1 := ??? :  mr1 := ??? :   mf2 := ??? :   mr2 := ??? :
>
```

• A Strong Recommendation. In your list enter specific values for mp and mf1, but then enter formulas for the quantities mr1, mf2, and mr2. These formulas should be valid for any values assigned to mp and mf1. Setting up your equations in this manner will greatly simplify your work in the next section.

Our procedure to numerically model the two stage rocket is remarkably simple: two applications of RocketMotion (one for each stage), with the combined data fed into VelocityPlot. Here are the details.

Use RocketMotion to determine the position and velocity of the rocket at the end of the burn period of the first stage assembly. Think carefully on the values you assign to the "total mass" m0 and the "fuel mass consumed" mf. Enter your commands below.

```
> m0 := ???; mf := ???;
> data1 := RocketMotion ??? :
```

Use VelocityPlot to construct a plot of the velocity and escape velocity data that you just produced with RocketMotion. Is the first stage able to make the rocket escape from the Earth's gravitational pull? Place your command and answer below.

```
> VelocityPlot ???
```

Answer:

We now turn to the second stage of the rocket. After the first stage burns out (i.e., consumes all its fuel), it drops off and is no longer of relevance to our problem. At that moment the second stage begins its burn (for tBurn = 30 sec). Use RocketMotion to determine the position and velocity of the rocket at the end of the burn period for the second stage assembly. Important Hints!!:

• Hint: The values for the final time, position, and velocity are contained in the list output from RocketMotion. Just access then with the command data1[4], data1[5], data1[6], for tf, sf, vf respectively.

Think carefully on the values assigned to the "total mass" m0 and the "fuel mass consumed" mf.

You will need to assign new values to some of the RocketMotion options, e.g., InitialPosition, InitialVelocity, and InitialTime.

Enter your commands below.

```
> m0 := ??? ;   mf := ???;
> data2 := RocketMotion ??
> VelocityPlot(data1,data2);
>
```

Check your plot. If you have done things correctly, the time span on the independent axis should be 60 seconds, and you should have only one blue line and one red line. Also, although there might be a sharp corner in your red curve at t=30 seconds, there should certainly not be an actual break in the curve at this point. Should your plot be inaccurate in any of these ways, go back and look at the three Important Hints, because you probably did not heed all of them. Make corrections (if necessary) in your RocketMotion commands above.

Will the spent second stage and the payload escape the earth's gravity? Explain. Place your answer below.
Answer:

¤ 4. The Two Stage Rocket: the General Case.

The commands in the previous section that we used for mf1 = 7000 can be generalized to apply to any allowable value of mf1. An important step in making such a generalization is to eliminate the need for the user to "intervene" after the computation has begun. In the previous section we did have to "intervene": we noted the values of s and v at the end of the first stage burn, and then we entered them in the RocketMotion command that modeled the second stage burn.

Set up a collection of Maple commands that will take any value of mf1, compute the velocity and height after the first stage has burned out, and then automatically compute the velocity and height after the second stage has burned out. Finish with an appropriate plotting command. All this can be accomplished by "glueing together" four groups of commands which you developed earlier:

The commands assigning values to the mass constants mp, mf1, mr1, mf2, and mr2.

The RocketMotion command which models the second stage burn. At least some of the values captured from the first stage burn need to be fed into this second RocketMotion command in an appropriate manner.

A VelocityPlot command which will graph the velocity and escape velocity data sets generated by the two RocketMotion commands.

Enter your commands below.

>
>

Using the Maple routine which you just constructed, determine if there are any values of the fuel load mf1 for the first stage that will result in the payload escaping the Earth's gravitational pull. If successful, record your value for mf1 below. If unsuccessful, explain why. Enter your commands and answers below.

Answer:
☐

LAB 14: World Population Growth

□ NAMES:

□ PURPOSE.
Once again we study the growth of a population, this time the human population of the world. We first compare the historical data on world population with the Malthusian or "natural growth" model, dP/dt = k*P, then with the von Foerster or "coalition" model, dP/dt = k*P^(1+r). Our main purpose is to determine whether either of these models can adequately fit the general trend of world population growth and, if so, to determine appropriate values for the parameters, find an approximate formula for the population function, and explore the consequences of the historical trend.

In order to carry out this ambitious program, it is useful to develop an efficient method for the numerical approximation of derivatives--symmetric difference quotients--as well as recall some commands for the manipulation of lists in Maple.

□ PREPARATION.
¤ This is a long and sophisticated laboratory--you will not finish it within the allotted time span unless you read these instructions carefully before the start of the laboratory session. In particular, be sure to work out and write down the answers to the substantive "content" questions asked in the Preliminaries section. During the lab session you will then be able to check your work against the answers found by activating the next command line.

¤ Study the first section of Chapter 7's Lab Reading, pages 454 - 456 in Calculus: Modeling and Application on the coalition model. In particular, work out Problem 1; your solution will be part of your report. You may find Problem 1 easier if you have already done Exploration Activity 2, starting on page 410.

¤ Review Section 2.6 in Calculus: Modeling and Application on semilog and loglog plots. Such plots will play a key roll in this laboratory.

□ LABORATORY REPORT--General Instructions.
Note: Follow your instructor's instructions if they differ from those given here.

You are to write a self-contained, well-organized exposition and analysis of the issues raised in this laboratory. In particular, it should incorporate the answers to all the important questions raised during the lab session.

Two submissions will be made for this report: an initial submission, and a final revision. Comments made on the initial submission will help you prepare the revision, and only the revision will receive a grade. Moreover, the grade will be based solely on the written report--the instructor will look at other parts of the worksheet only in unusual situations. You should therefore consider the entries you make in the Preliminaries and Project sections to be notes to help when later writing your report.

More specific instructions on the content of the report will be given at the end of the worksheet.

□ Maple Packages. Activate the following line.

```
> with(lab14);
> # If an Error is an output of the above line, see your lab instructor.
```

□ PRELIMINARIES.

¤ 1. Maple and Lists.

It's time to review how Maple handles lists. The following observations should be helpful when working with data.

• Basic List Manipulation.

Here are some data sets to use in the examples which follow. Enter the following line.

```
> list1 := [ c1, c2, c3, c4, c5, c6, c7 ];
> list2 := [ [a1,b1], [a2,b2], [a3,b3], [a4,b4], [a5,b5] ];
```

Now Enter each of the following lines. The output should tell you what each command does. Enter each of the following lines and consider the output.

```
> n1 := nops(list1);
> n2 := nops(list2);
```

• Maple Tip. The Maple function nops returns the number of operands in the expression. For a list or a set, the number is the number of elements in it.

```
> sublist1 :=[ list1[2..5] ];
> sublist2 := [ list2[1..n2-1] ];
> sum1 := sum(list1[k], k=1..n1); # sum of elements of list1.
> sum2 := sum(list2[k], k=1..n2);
```

Using these basic commands we can construct a new Maple command zapFirstAndLast which will take an arbitrary list list0 and remove the first and the last elements. You will need this later in the laboratory project. Test this command on list1 and list2. Enter the following commands.

```
> list0 := [d1, d2, d3, d4, d5, d6, d7];

> zapFirstAndLast := list -> [ list[2..nops(list)-1] ]; # This is needed later.

> zapFirstAndLast(list0);
> zapFirstAndLast(list1);
> zapFirstAndLast(list2);
```

• Zip utilities contained in the Lab 14 Procedures are:
Zip, unZip, unZip1, and unZip2

Study the effects of these commands by Entering the following examples. Enter the following lines, one at a time, and examine the output.

```
> list1 := [ c1, c2, c3, c4, c5, c6, c7 ];
> list2 := [ [a1,b1], [a2,b2], [a3,b3], [a4,b4], [a5,b5] ];

> unZip1([ [2,5] ]);

> unZip2([ [2,5] ]);

> aList := unZip1(list2);

> bList := unZip2(list2);

> aAndbLists := unZip(list2);

> Zip([a1, a2, a3, a4, a5], [b1, b2, b3, b4, b5]);

> Zip(aList, bList);
```

Here are some data sets to use in the examples which follow. Enter the next line.

```
> list3 := [ c1, c2, c3, c4, c5 ];
> list4 := [ [a1,b1], [a2,b2], [a3,b3], [a4,b4], [a5,b5] ];
```

Optional. On the basis of what you have just seen in the previous examples, write one or more Maple statements which will take the first element ai in each point in the list list4 and combine it with the element ci from list3, forming a new list5, consisting of the points [ai,ci]. Enter your commands here.

Why we have used the names Zip and unZip for these operations? Place your answer below.
Answer:

```
> Ans(1);
```

¤ 2. Symmetric Difference Quotients and Derivatives from Data.

Suppose s is a quantity which depends on a variable t, and we have measurements for s at the t-values t1<t2<t3. How could we numerically estimate the derivative of s at t2?

Later in this laboratory we will need to estimate derivatives of world population with respect to time. That is why we need accurate methods for approximating derivative values from data.

Our first thought, since we are working with data at a number of discrete t values, would be to approximate the derivative at t=t2 by the difference quotient

Deltas/Deltat = (s(t3)-s(t2))/(t3 - t2)

We claim, however, that a better estimate to the derivative is generally obtained by using what we shall call the symmetric difference quotient

(Deltas/Deltat)sym = (s(t3)-s(t1))/(t3 - t1)

As you can see, to approximate the derivative at t2 via symmetric difference quotient (sdq) we use the function points to either side of t2, i.e., (t1, s(t1)) and (t3, s(t3)).

To convince you that, in most cases, the symmetric difference quotient is indeed the better approximation to the derivative, we present some graphical evidence. Enter the following commands.

```
> func := t -> t^3;
> t2 := 1.0; run := .5;
> sdqPlotter(func, t2, run);
>
```

Explanation: The function func(t) = t^3 is plotted for the interval 0.5 <= t <= 1.5 . Three points are shown as small circles, corresponding to t1 = 0.5, t2 = 1.0, and t3 = 1.5. Delete the incorrect choices for each question concerning the center point t2.

The difference quotient is the slope of the
(red, blue, or magenta) line.
The symmetric difference quotient is the slope of the
(red, blue, or magenta) line.
The derivative is the slope of the
(red, blue, or magenta) line.
The derivative is best approximated by which quotient?
("regular" or symmetric)

Consider the case where we have a data set
[[t1,s1],[t2,s2],[t3,s3],...,[tn,sn]]
and we need to approximate the derivative ds/dt at the data points. As discussed above, we estimate the derivative of s=s(t) at each tk by a symmetric difference quotient, which now takes the form:

(Ds/Dt)sym = (s(k+1)-s(k-1))/(t(k+1) - t(k-1))

for k = 2, 3, ..., n-1 . Here n is the index of the last data point. (We do not estimate the derivative at the end points, t1 and tn, since we have no values for s(-1) nor s(n+1).)

To make sure that you understand this procedure, compute (by hand or with the computer) the first possible sdq for

worldData=[[1000, 200], [1650, 545], [1750, 728], [1800, 906],....].

Do your computation on paper, then check your answer by activating the line below.

```
> Ans(2);
```

For your convenience, we have a command sdq which computes the symmetric difference quotients for a data list. To find out more about sdq use the ? command. Enter your information request command below.

Apply sdq to the following list. Do you obtain what you expect? Complete and Enter to following commands.

```
> list5 := [ [0,0], [.5,.6], [1,2], [1.5,2.5], [2,3.2], [2.5,3.5] ];
> plot(list5, style=POINT, symbol=CIRCLE);
> sdq(???);
> Ans(3);
```

¤ 3. The Malthusian or "Natural Growth" Model.

A population which exhibits Malthusian growth satisfies the differential equation dP/dt=k*P for some constant k. Such a population will be modeled by what sort of function? Place your answer on paper, then activate the line below.

```
> Ans(4);
```

Hence what would be a method for examining world population data to determine if it exhibits Malthusian growth? (Hint: an important graphical technique from earlier in the course.) Place your answer on paper, then activate the line below.

```
> Ans(5);
```

The first thing we will do in this lab is to apply such a test to world population data.

¤ 4. The von Foerster or "Coalition" Model.

The von Foerster argument claims that the differential equation modeling the growth of the world population P as a function of time t might have the form

dP/dt = k*P^(1+r)

where r and k are positive constants. Using separation of variables we can show that any solution of this equation must be of the form

```
> ( 1/(r*k*(T-t)))^(1/r);
```

$$\left(\frac{1}{r\ k\ (T-t)}\right)^{\left(\frac{1}{r}\right)}$$

Most of the lab will be centered on attempts to find the best values of the constants r, k, and T to fit this model to the actual world population data.

We will first estimate values for r and k. Then, independent of our r and k values, we will estimate a T value by using a relationship between log(p) and log(T-t) that we will now establish.

If we take a log of both sides of the equation for P given above we obtain what new equation? Write your answer on paper, then activate the next line.

```
> Ans(6);
```

If r and k are fixed, then what sort of a relationship does this give between the variables log(p) and log(T-t)? (i.e., quadratic? exponential? linear?) Write your answer on paper, then activate the next line.

```
> Ans(7);
```

We will estimate a T value for world population growth by using a LogLog plot between the two quantities T-p and p.

□ PROJECT.

¤ 1. Was Malthus Right?

Here is the world population data as best as we know it. The first number in each data pair is the year, and the second number is the population in millions. Enter the following command.

```
> worldData:=[ [1000, 200], [1650, 545], [1750, 728], [1800, 906], [1850,1171],
> [1900,1608], [1920,1834], [1930,2070], [1940,2295], [1950,2517], [1960,3005],
> [1976,4000], [1987,5000], [1991,5400] ];
```

The issue we consider in this section is whether or not the world population data exhibits Malthusian growth. As a start, plot the data in the "ordinary" way, using plot. Does it look like it exhibits Malthusian growth? Enter your plotting command and observations below.

```
> plot( ???);
```

Answer:

```
> Ans(8);
```

Eyeballing the graph will tell you some things, but it cannot prove Malthusian growth--lots of curves have this same basic increasing shape but yet are far from true exponentials. However... there is a graphical way to answer this question. What is that method? Place your answer below.
Answer:

For your convenience we provide commands which produce semilog and loglog data from a given data set: semiLogList and logLogList. This data can then be plotted with the plot command. Use the "?" to obtain more information about these commands, and then choose the appropriate one to apply to worldData. Enter the appropriate plotting command below.

• Maple Tip. Adding the option labels=[` year`, `p (millions)`] to your plotting command improves the clarity of the resulting graph.

```
> semiLogData := semiLogList(worldData);
> plot( ??? );
>
```

On the basis of the plot you have just produced, do you believe that Malthus was right, i.e., has world population been growing according to his "natural growth" model? And if not, were his growth predictions too large or too small? Place your answer below.
Answer:

¤ 2. Was von Foerster Right?--An Initial Check.

We now ask a new question: can world population be approximated by the coalition model, i.e., can we find constants r and k such that the differential equation

$$dP/dt = k*P^{(1+r)}$$

will fit our world population data? And if so, what conclusions can we draw from the model?

In order for the coalition model to be valid, we see that

• the derivative dp/dt will need to be proportional to a power of the population p.

Thus, unlike the "natural growth" model considered above, we will compare dp/dt with p, not p with t. We have a graphical procedure for determining whether or not two quantities obey a power relationship--what is it? Place your answer below.
Answer:

In order to carry out this test we must have a new data set--call it derivData--consisting of pairs of the form [p,dp/dt]. We construct this from the original data set worldData in the following way. First we obtain the p values by "unzipping" them out of worldData. Enter the following command.

```
> pValues := unZip2(worldData);
>
```

Next we obtain numerical approximations to the corresponding derivatives dp/dt by applying sdq to worldData. (The command sdq was discussed in the Preliminaries.) Complete and Enter the following command.

```
> sdqValues := sdq(???);
>
```

Finally derivData is obtained by "zipping" together pValues and sdqValues, then throwing away the useless first and last points. Enter the following commands.

```
> derivData0 := Zip(pValues, sdqValues);

> derivData := zapFirstAndLast(derivData0);
>
```

Recall that we wish to carry out a graphical test for determining if dp/dt is proportional to a power of p. This test is now easy to execute given the data set derivData and the specialized plotting commands described in the previous section. Execute this test, calling your plot testPlot. Enter the appropriate plotting command below.

```
> logLogDerData := logLogList(derivData);
> testPlot := ???
>
```

Does this plot make you more skeptical or less skeptical about the validity of the coalition model for modeling world population growth? Place your answer below.
Answer:

¤ 3. Determining the "Coalition" Model Parameters.

You will need testPlot, pValues, and sdqValues from the previous section. To check that these have been properly defined, Enter the following command.

```
> plots[display](testPlot);
> pValues;
> sdqValues;
```

Entering these statements should have produced the plot from the end of the previous section, followed by the two lists pValues and sdqValues. If any of these actions did not occur, then go back to the previous sections and reEnter any command that is relevant to testPlot, pValues, and sdqValues.

You can determine values for the "coalition model" constants r and k by fitting a line through the loglog plot of derivData that was constructed in the previous section (this was labeled testPlot). We have set up some commands to make this process easy to carry out. Choose initial values for the slope and the y-intercept of the best modeling line (start with approximations from testPlot). Place these values on the first two lines below. Then Enter the line. The line will be drawn on your plot so that you can judge the accuracy of your slope and y-intercept values. Refine the values and Enter the commands again. Continue in this way until you get a good fit to your plot. Complete and Enter the commands below--multiple times if necessary.

```
> m := ??? ;   b := ???;
> plots[display]([testPlot, plot(m*t+b, t=6..9, color=magenta)],labels=[`ln(p)`,`ln(dp/dt)`]);
>
```

Assuming that you have a good linear fit to your data, convert your line data into a function called line(x) by using the usual m*x + b form of a line. Enter the following definition of the line.

```
> line := x -> m*x + b;
>
```

Hence an equation for the line you have obtained in testPlot is given by

y = line(x) = m*x + b,

where x and y are the independent and dependent variables in testPlot. What are x and y in terms of dp/dt and p, i.e.,

x = ???

y = ???

This is equivalent to asking what x and y stand for in the testPlot that you generated above. Place your answers above, then check by activating the following line.

```
> Ans(9);
>
```

Combining these formulas for x and y with the formula for y=line(x), we obtain an equation relating dp/dt and p. What is this equation? Write out your answer on paper, then check by activating the following line.

```
> Ans(10);
```

We can get rid of the extraneous logs by exponentiating both sides of this equation by base e. Applying the laws of logs and exponentials yields a simple formula for dp/dt in terms of p. Determine this formula! Write out your computations on paper, then check by activating the following line.

```
> Ans(11);
```

Referring back to the results of the line fitting process, what values have you computed for m and b? Enter them as Maple definition commands. Enter your values for m and b below.

```
> m := ??? ;  b := ???;
```

We now can compare our results with the original differential equation:

dp/dt = k*p^(1+r).

From this comparison we can read off the values of k and r in terms of m and b: Enter your statements below.

```
> k := evalf(???);  r := ???;
```

So we now have a model for the derivative of p as a function of p:

dp/dt = dpdt(p) = k*p^(1+r).

Let's enter the modeling function dpdt so that we can use it. Enter the following function definition statement.

```
> dpdt := p -> k*p^(1+r);
```

To check this modeling function (dp/dt as a function of p), let's plot it (an ordinary plot, not a loglog plot) along with our data derivData. The hope is that the graph of the modeling function will provide a good approximation to the scatterplot of derivData. For this we will use the plots[display] command.

```
> plot1 := plot(derivData, style=POINT,symbol=CIRCLE,color=red):
> plot2 := plot(dpdt, 500..5500, color=blue):
> plots[display]([plot1, plot2]);
```

How good a match do we have? Place your answer below.

Answer:

• An alternate derivation for k, given r.

Although we seem to have a good match between the derivative data and the modeling function, when we look at the original population data, we are going to be able to improve the fit by jiggling the r and k values a bit. For that reason it is important to have an alternate method of calculating values of k. In particular, given a value of r, we would like a formula that will give us a corresponding value of k.

To do this, recall the differential equation that we are dealing with:

dp/dt = k*p^(1+r).

Given a value of r, we can then use our data lists pValues and sdqValues to form the quotients

(dp/dt)/p^(1+r)

These numbers should all be approximately equal, the common value being k. So let's compute all these values, average them together, and call the new result kFunc(r). It's impressive that Maple can do all of these steps in a one line statement. Here it is. Enter the following function definition statement:

```
> n := nops(worldData);
> kFunc := r ->sum(sdqValues[i]/pValues[i]^(1+r), i=2..n-1)/(n-2);
```

Look carefully at this definition of kFunc, make sure that you understand what it is supposed to do and that it in fact does it. Test out kFunc(r) against the values of r and k which you found above. Enter the following command.

```
> r; k; kFunc(r);
```

¤ 4. Determining Doomsday.

Recall that the solution to the coalition growth model is of the form

P = (1/(r*k*(T-t)))^(1/r)

where r and k were computed in the previous section, and T is yet-to-be determined. We do, of course, have a standard method at this point for finding the value of T: since we "know" r and k, we can plug one of our population data points into the equation and solve for T.

There is, however, a drawback to this method: if we wish to "jiggle" the values of r and k, we would also have to recompute T each time. Fortunately there is another way to compute T, totally independent of values for r and k.

Recall our observation in Preliminaries that, given the "correct" value of T, there is a linear relationship between the variables log(p) and log(T-t).

Our search for the "correct" T can thus proceed as follows: given any value of T, we create a scatter plot of the data [log(T-t), log(p)] and examine whether or not the data falls roughly in a linear pattern. Do this for numerous T values and pick that value which gives "the best" line.

For convenience, we have put all these instructions together in one function: TPlotter. Use the "question mark" command to obtain more information about TPlotter, and then Enter the following "sample run" of this routine. Enter the following commands.

```
> ?TPlotter
> T:=1998; TPlotter(T);
```

Enter your guesses for T and study the resulting graphs. Devise a strategy for choosing successive T values that will make the scatter plot as straight as possible. Be sure to clearly indicate the value of T that you finally settle upon. Enter your final value of T below:

T = ???

¤ 5. Was von Foerster Right?--Conclusion.

You will need r, k, T, and kFunc from the previous sections. To check that they have been properly defined, Enter the following commands.

```
> r; k; T; kFunc(r0); # kFunc was defined at the end of Section 3
>
```

Entering these commands should have produced (1) a list of three numbers (the values determined for r, k, and T), and (2) a complicated formula for kFunc(r0). If one or both of these did not happen, then go back and Enter the necessary statements from the previous sections.

*****

We now have estimates for all three of the constants that appear in the coalition growth model solution: r, k, and T. Hence we can plug them into the coalition solution

P = (1/(r*k*(T-t)))^(1/r) = (r*k*(T-t))^(-1/r)

plot the function, and compare it with the actual population data in worldData.

We first need to define our model as a function p=pModel(t). We do this by converting the coalition solution, just given above, into a Maple definition for pModel(t). Enter the following Maple definition for pModel:

```
> pModel := t -> (r*k*(T-t))^(-1./r);
```

$$\text{pModel} := t \to (r\ k\ (T - t))^{\left(-1.\ \frac{1}{r}\right)}$$

We now can use the Maple command plots[display] to plot worldData along with the graph of the function p=pModel(t). Enter the commands below.

```
> plot1 := plot(worldData, ??? ):
> plot2 := plot(pModel, ??? , color=blue):
> plots[display]([plot1, plot2], labels=[`t`,`p`]);
```

You should have a reasonably good fit to the curve...but we can do even better! We modify our model function and the plotter to allow for different values of r. Now we can "jiggle" the r parameter to see if we can achieve a better fit with the data. The k value will automatically change to correspond to new values of r because we will use the kFunc function defined at the end of Section 3. No change is needed in the T value because we obtained its value without reference to r and k. Enter the following Maple command.

```
> pModel2 := (t,r) -> (r*kFunc(r)*(T-t))^(-1./r);
```

The Maple command modelRefiner has been programmed below so that you can change r and see the resultant approximation of the worldData plot with the pModel2 plot. To get the desired plots you simply activate the next region to define modelRefiner and then in the following region, assign a numerical values for r and activate the region. Find the r value that gives the best fit to the data. Indicate your final choice for r below.

```
> modelRefiner := proc(r)
>  local plot1, plot2, plot3, rs, ks, Ts;
>  plot1 := plot(worldData, style=POINT,symbol=CIRCLE,color=red):
>  plot2 := plot(pModel2(t,r), t=1000..2100, color=blue):
>  rs:= convert(r, string):      # Used to insert info into graph.
>  ks := convert(kFunc(r), string):
>  Ts := convert(T, string):
>  plot3:=plots[textplot]([[1500,12000,`r=`.rs],[1500,10000,`k=`.ks],
>           [1500,8000,`T=`.Ts]]):
> plots[display]([plot1,plot2,plot3], view=[1000..2100, 0..15000], labels=[`t`,`p`]);
> end:
> r := ???:  T:= T:      # You can change T too if you like.
> modelRefiner(r);
```

Place your final choice for r below and copy and paste a copy of your best plot.

Answer: r = ???

Your best plot:

¤ The Residual. (Optional)

We can fine-tune our fit even more by looking at the residual, the collection of differences between the actual data and the modeling function. Thus, if
worldData = [[t1,p1], [t2,p2], ...], then
residual = [[t1, p1-pModel(t1)], [t2, p2-pModel(t2)], ...]

A "good" fit will result in a "random" looking residual centered on the t-axis; a "bad" fit will result in some "non-random" pattern emerging in the residual. The following two lines compute and plot the residual of worldData as modeled by pModel(t). The residual function has been supplied with this lab. Enter values for r and T and activate the following region.

```
> r:=???:  k:=kFunc(r):  T:=???:
> plot(residual(r), style=POINT, symbol=CIRCLE, color=red);
```

(Optional) If your residual does not look random, then change the value of r (and possibly T) to attempt to obtain a "better" plot.

When you have obtained as good a residual as possible, make one last plot of the data along with the graph of the modeling function using ModelRefiner. Enter the following command.

```
> modelRefiner(r);
```

Well, was von Foerster right?!? Before you answer based only on the data seen above, let's use the coalition model just obtained to predict some populations in the future. You have only to apply the function pModel to the desired years. However, you must make sure that r, k and T are set at their optimal values, as determined above. Enter your computations below.

```
> t := 't';
> plot(pModel(t),t=???);
```

What are the meanings of these predictions? Are they realistic? Do these predictions invalidate the model? Are you worried? Why do we refer to T as Doomsday? Place your answers below.

Answer:

□ LABORATORY REPORT--Specific Instructions.
Note: Follow your instructor's instructions if they differ from those given here.

Write a self-contained, well-organized exposition of the issues raised in this lab. In particular, your report should not be a simple step-by-step restatement of the project--that shows little recognition nor analysis of the central ideas of the lab. Instead, as with any good expository paper, you should begin with a clear understanding of the principal ideas you wish to communicate, and then decide on a structure for your report that will best convey these ideas. You can then discuss specific portions of thc lab project in the appropriate parts of the report. An example of a "generic" organization that can be used for a laboratory report is to divide your essay into four sections:

(1) Purpose and goals.
(2) Procedures.
(3) Results.
(4) Analysis.

For this lab, obtaining a clear and complete "Procedures" section will probably require the most careful thought. In particular, you will need to carefully explain how you determined your best values for r, k, and T in the coalition model. This will require an explanation as to why so much analysis was done with the data set derivData (consisting of points [p,dp/dt]) instead of the given data set worldData (consisting of points [t,p]). (Consider alternate methods: can you devise a method to determine r, k, and T simply by analyzing the [t,p] data in worldData? This exercise should make you realize why it was necessary to use the [p,dp/dt] data.)

You should also include the solution to the first portion of Problem 1 from Chapter 7 Lab Reading in Calculus: Modeling and Application. The "Analysis" section will also require some careful thought. This should include a discussion of the reasonableness of the coalition model and the significance of the quantity T. How reasonable is the value which you found for T during the lab? Does this invalidate the coalition model, or make it unrealistic? Give a short discussion of the philosophical and practical implications of what you have determined from the historical population data. Remember to keep your audience in mind: you should assume your reader is comfortable with the concepts of calculus, but is not familiar with the particular material of this laboratory.

Place your final report below.

□

LAB 15: Chaotic Dynamics

□ NAMES:

□ PURPOSE.
The primary purpose of this lab is to discover Feigenbaum's constant, a "universal constant" associated with bifurcation points that appears frequently in many apparently different contexts associated with chaotic dynamics. (This constant has no known connection with Pi, e, or any of the more familiar universal constants.) A secondary purpose is to discover an attracting cycle of a logistic iteration of an order that cannot arise from bifurcation.

□ PREPARATION.
¤ Study Section 7.3 of Calculus: Modeling and Application, pages 426 - 443, and do all the exercises embedded in the section. In particular, carefully read the subsection "Convergent, Periodic, and Chaotic Iterations," which describes the web diagram. Bring this material to the lab.

¤ Obtain a paper copy of the Chaotic Dynamics Laboratory Results Summary Sheet. The summary sheet will be found at the end of this file.

□ ACKNOWLEDGMENTS.
Much of the code used in this lab is based on programs given in Chapter 7 of Theodore Gray and Jerry Glynn, "Exploring Mathematics with Mathematica," Addison - Wesley Publishing Company, 1991.

□ LABORATORY REPORT--General Instructions.
Note: Follow your instructor's instructions if they differ from those given here.

You are not required to submit a formal report. Instead, you are requested to turn in an electronic copy of the worksheet and a copy of the Chaotic Dynamics Laboratory Results Summary Sheet, with the questions answered and the tables completed. (A copy of the Summary Sheet will be found at the end of this worksheet.) Only one submission will be made for this laboratory. Be sure to express your answers in complete and understandable sentences.

□ Maple Procedures. Please activate the following line.

```
> with(lab15);
> # If an Error is the output of the above line, see your lab instructor.
```

The Maple procedures loaded, define a number of new commands among which are: webList, webDiagram, movieWebLa, OrbitPlot, CycleFixedPoint, LengthList, and LogLengthPlot. All these commands are discussed later in the worksheet.

□ WEB DIAGRAM PROJECT.

¤ Web diagrams.

We will be studying the famous logistic map

y = F(x) = lambda*x*(1-x),

defined for x in [0, 1], where lambda is a fixed real number between 0 and 4. To understand this map let's start by looking at its graph. Activate the following region for various values of lam between 0 and 4.

```
> lam := 3.5:
> plot(lam*x*(1-x), x=0..1, y=0..1);
```

Question. For what value of x does F(x) attain its maximum value? For a given value of lambda, what is that maximum height?
Answers:

To better understand F we will start by looking at the sequence of iterates with initial point x0 for a fixed lambda,

x0, x1=F(x0), x2=F(x1), x3=F(x2), ... , x(n+1)=F(xn), ..., etc.

To facilitate the study of this sequence of iterates, the Lab 15 Procedures define a command called webList which lists successive iterates of a point such as listed above.

- webList. Activate the following line.

```
> webList(2.9, .1, 0, 4);
```

First argument: 2.9 = lambda value of F,
Second argument: .1 = initial point x0 of the sequence.
Output: .261 = F(.1), .5593491 = F(.261), etc.
Third argument 0 means no values in the sequence were skipped.
Fourth argument 4 means that 4 values are listed.

We now change the third argument to a 2. Activate the following line and compare the output with the previous output.

```
> webList(2.9, .1, 2, 4);
```

The 2 in the third argument means that the first two values in the sequence of iterates have been skipped and the 4 in the fourth argument means that four values are shown after the two that are skipped. Activate the following command for more information on webList.

```
> ?webList
```

Exercise. Use the webList command to evaluate what happens to successive iterates of x0 = 0.1 for lambda = 2.5 . Do not skip any values, that is start with x0 = .1 which means the third argument "skip" should equal 0. Do the values tend to converge? To What?

```
> webList(2.5, 0.1, 0 , 10);
>
```

Answer:

Exercise. Do the same for lambda = 3.1 . Do the values tend to converge, settle down to alternating between two or more numbers, or does something else happen?

```
> webList(3.1, 0.1, 0, ???);
>
```

Answer:

Question. What difference does it seem to make if you go back and try an initial starting point different than 0.1 ?
Answer:

Lab 15 also has a command called

- webDiagram.

The command webDiagram provides a graphical version of webList. Refer to the subsection "Convergent, Periodic, and Chaotic Iterations" in Section 7.3 of Calculus: Modeling and Application for a definition of web diagrams. The calling sequence is

webDiagram(lam, x0, skip, show)

It uses the same arguments as webList.

Enter the following line to learn more about the command.

```
> ?webDiagram
```

Use the webDiagram command to show what happens when we take lambda=3.5, start with x0=0.1, skip no cycles, and show just one "cycle". Enter your webDiagram command below.

```
> webDiagram(3.5, .1, ???, ???);
```

The parabola you see in this plot is the graph of the equation

y = F(x) = lambda*x*(1-x), for lambda = 3.5 .

The red line segment goes from the starting point [0.1, 0.1] up to the parabola to the point (0.1, 0.315). In what way does the function F(x) relate these two numbers, 0.1 and 0.315? (Hint: what is the meaning of a web diagram?) Place your answer below.
Answer:

Using the same lambda and x0 values, devise a command to show two full cycles, not just one--thus we will continue to skip no cycles, and we will show the two cycles that are generated. Enter your command below.

```
> webDiagram(3.5, .1, 0, ???);
```

The parabola you see is still the graph of the equation

y=F(x)=lambda*x*(1-x),

but now illustrates an additional iteration of the function. The two new red line segments go from the point (0.1, 0.315) horizontally to (0.315, 0.315) and then up to the parabola to (0.315, 0.7552125). In what way does the function F(x) relate these two numbers, 0.315 and 0.7552125? How are .1 and 0.7552125 related? Place your answers below.

Answer:

How does a web diagram relate to the corresponding web list?

Answer:

• Maple Tip. Recall that in Maple, the command

plot([[x0, x1], [x1, x2], [x2, x3], ...])

yields an interval from [x0, x1] to [x1, x2], an interval from [x1, x2] to [x2, x3], etc.

Problem 1. Let A = [x0, x1, x2, ...] be the sequence of iterates of an initial point x0 with the function F, that is, x1=F(x0), x2=F(x1), etc. Let S be the sequence obtained by replacing each xn in the above sequence by the pair of pairs [xn, xn], [xn, F(xn)]. That is,

S = [[x0, x0], [x0, F(x0)], [x1, x1], [x1, F(x1)], [x2, x2],[x2, F(x2)], ...]

Explain why it should be the case that if A is the output of a webList command, then the output of the corresponding webDiagram should be the plot of F combined with the plot of S.

Answer:

• Animation tool--vary the number of cycles.

We can animate the web diagrams by using the seq command to build several of the web diagrams, and then display them with the plots[display] command with the option "insequence=true" command. Activate the following command region.

```
> frames := [seq(webDiagram(3.5, 0.1, 0, k), k=1..8)]:
> plots[display](frames, insequence=true);
>
```

We have a built-in procedure movieWebN that does this in a convenient way. Briefly,

movieWebN(lam, x0, skip, N)

where "lam" = lambda value, "x0" is the initial point, "skip" = number of cycles skipped, and "N" = number of plots in the animation showing additional cycles beyond those skipped.

Enter the following command to learn more about it.

```
> ?movieWebN
```

Now try this new animation command by activating the following line.

```
> movieWebN(3.5, 0.1, 0, 8);
```

Continuing with lambda=3.5, now generate a few cycles (say about 30) and make a web diagram where the first few cycles (say 15) are "skipped." In this way the "confusion" of the initial cycles is removed and it can be seen if a limit cycle is achieved. Some patience may be required while waiting for the computer to do its work. You can speed it up by lowering the number of cycles (last argument).

```
> ???
```

Now do it again skipping the first 20, then 25, and finally 30. Enter your commands below. Also use to corresponding webList command with the corresponding data.

```
> ???
```

You should see in the last plot (skip = 30) what appears to be a limit cycle. If the lines representing the cycle intersect the line y=x in n points, then we say we have found an n-cycle. What sort of cycle does it appear we have found?
Answer:

- Animation tool--vary lambda.

Lab 15 Procedure define the animation command movieWebLa. Briefly,

movieWebLa (laMin, laMax, Dx, x0, skip, show),

which provides movies of the webDiagram procedure as lambda varies from "laMin" to "laMax" in steps of "Dx", for the initial point "x0" where "skip" is as before and "show" = number of cycles to show after starting to graph.
(The number of plots in the animation will be (laMax-laMin)/Dx.)

Using movieWebLa, vary lambda between 2.9 and 4.0 in steps of 0.1 where x0=0.1, skip=30, show=40. Enter your commands below.

```
> ???
```

Between what lambda values do we first begin our descent into chaos? Place your answer below.

Use webDiagram or movieWebLa to track down a value of lambda which gives an 8-cycle limit. (Hint: From the previous results, for what range of lambda values do you think it most likely that an 8-cycle will be found?) Enter your computations below.

```
>
```

Answer:

□ BIFURCATION PROJECT.

¤ 1. The bifurcation diagram.

We can visualize the eventual behavior of iteration for increasing values of lambda with a "bifurcation diagram." A bifurcation diagram contains, for each particular value of lambda, a plot of the x-values to which the iteration leads. Thus, for lambda < 3, we plot only one x-value. For lambda > 3, we begin plotting multiple x-values for each lambda. A bifurcation diagram locates lambda on the horizontal axis and x on the vertical axis because the x-values that are plotted depend on the value of lambda. Maple generates bifurcation diagrams for a specified range of lambda-values. For each lambda, the beginning iterations (known as "transients") are generated but not plotted, allowing the iteration process to settle into a well-defined pattern. The software then lots a certain number of subsequent iterations. Which may or may not be different x-values.

In short, the bifurcation diagram for the logistic iteration

F(x) = lambda*x*(1-x)

plots the limit cycle values as a function of lambda.

You will now use OrbitPlot to construct a bifurcation diagram. Briefly,

OrbitPlot(LaMin, LaMax, Num, x0, skip, show)

where Num is the number of evenly placed lambda values between LaMin and LaMax. Obtain more information about this new command by using the "?" command. Enter your command below.

```
> ?OrbitPlot
```

Set the lambda limits to be 2.9 and 4.0, request 100 lambda values, a starting value of 0.4, skip 200 cycles and show 30. Enter your plotting command below.

```
> ???
>
```

How many points are in the limit cycle for lambda=3.55? Place your answer below.
Answer:

Zooming: To zoom in on a window of the Orbit plot output, use the mouse to find all four corners, and then input those into the OrbitPlot command with the use of the view=[x0..xf, y0..yf] option.

e.g. OrbitPlot(2.9, 3.1, 100, 0.4, 200, 30, view=[2.9..3.1, 0.6..0.8]);

¤ 2. Self-similarity.

Refer to the first bifurcation diagram that you obtained in Section 2. A curve comes in from the left and then splits into two new curves at lambda = 3.0 (we thus call lambda = 3.0 a bifurcation point). The diagram now begins to look something like a big tooth held sideways, with two smaller "teeth" nestled at its base (on the right). Choose one of these smaller "teeth" and zoom in on it using OrbitPlot. Enter your command below.

```
> ???
```

What relationship do you see between the full bifurcation diagram and the portion of the diagram that you just produced? Might you have a good name for this property? Place your answer below.

Are there other parts of the bifurcation diagram that might exhibit a similar behavior? Zoom in on one or two likely candidates and report what you find. Enter your commands and answers below.

```
>
```

Answer:

¤ 3. The discrete case verses continuous case.

The continuous logistic model is given by the differential equation

dp/dt = k*p*(1-p)

The corresponding discrete logistic model is given the by difference equation

Deltap/Deltat = k*p*(1-p)

To relate these two models to the bifurcation diagram, we perform the same scaling changes that are described in "A Rescaling of Convenience" in Section 7.3 of Calculus: Modeling and Application.

Replace p with x*(k+1)/k
and k with lambda - 1.

When you make these changes, the continuous logistic model becomes
dx/dt = lambda*x*((1-1/lambda) - x), (*)

while the discrete logistic model becomes

Deltax/Deltat = lambda*x*((1-1/lambda) - x). (**)

Letting Deltat=1 and Deltax=x(n+1)-x(n), the discrete model (**) yields the following recursion relation:
x(n+1) = lambda*x(n)*(1 - x(n)).
For each lambda, the values of x in the limiting cycle for this recursion relation are what get plotted in the bifurcation diagram. Now compare this with the continuous case. Equation (*) models a situation where 1-1/lambda represents the maximum sustainable value of the quantity x. In such a case we know that the limit of x as t becomes large is the maximum sustainable value, i.e., 1-1/lambda. Hence, in the continuous case, for each lambda there is just one limiting value, 1-1/lambda, while for the discrete case we can have multiple "limiting" values, leading to limiting cycles and ultimately the bifurcation diagram.

So how does the single limiting value in the continuous case compare with the limiting cycle values of the discrete case? This is easy to examine: we simply plot the graph of 1-1/lambda (the continuous case) on top of the bifurcation diagram (the discrete case). Enter the following command.

```
> lam := 'lam':
> plots[display]( [ plot(1-1/lam,lam=2..4,view=[2..4,0..1], color=red),
>                    OrbitPlot(2,4,200,0.3,100,30) ] );
>
```

Explain the relationship that you see between the continuous limiting value for a given lambda and the discrete limiting cycle for that same lambda. For what values of lambda does the discrete model behave in a way that is similar to continuous model? For what values is just the opposite true? Place your answer below.

Answer:

¤ 4. Estimating bifurcation points.

A primary objective of the lab is to determine the lambda coordinates of the bifurcation points, i.e., the values of lambda at which the single limit value for lambda "splits" into two values, then at which each of those "branches" splits again to make four branches, and so on. You should be able to read off approximate coordinates for a number of the bifurcation points from the diagrams you have already constructed, and you can construct further diagrams using OrbitPlot. However, we will provide you with additional tools to help in determining the bifurcation points.

The additional tools are the commands CycleFixedPoint, LengthList, and LogLengthPlot, and NestPlot. Obtain more information about these commands in the usual way. Place your "information searching" commands below.

```
>
```

Tip. CycleFixedPoint and LengthList have an optional argument called Precision and its default value is 10. For many computations a Precision=7, 8, or 9 would be adequate and would compute much faster.

Tip. Use LengthList and LogLengthPlot to approximate the bifurcation points, then "zoom in" for better approximations and use CycleFixedPoint to refine results.

```
> B := LengthList(2.9,3.60,40,0.4,Precision=8);
> LogLengthPlot(B);
> CycleFixedPoint(f, .4, 3.02, MaxIteration=17000);
> NestPlot(2,3.1);
```

Use the NestPlot command to examine the graph of f and f^2 on either side of the lambda value corresponding to the first bifurcation point.

Question. Describe in terms of the graphs what happens at the bifurcation points.
Answer:

Question. Does the number of iterations CycleFixedPoint uses to find a limit cycle depend on how close lambda is the a bifurcation point? If so, why?
Answer:

You now have powerful tools with which to estimate the bifurcation values. Using any of the tools at your disposal, estimate as many bifurcation values as possible, and estimate them as accurately as possible. Going up to the eighth bifurcation point (from 128 to 256) would be ideal. The quality of your lab will be judged in part by how successful you are with this exercise. You will need good approximations of the bifurcation points in order to estimate Feigenbaums's constant in the next section. Record the values that you find on the Chaotic Dynamics Laboratory Results Summary Sheet. Place your commands below.

```
>
>
```

Maple Tip. Suppose a finite orbit is reported that you are suspicious about (e.g., it might have a "funny" cycle length). Then reducing the size of Precision may confirm or deny your concern. (Reducing the size of this option makes the standard used for declaring the existence of a finite limit cycle more stringent, which might eliminate the suspicious lambda value.)

¤ 5. Estimating Feigenbaum's constant.

Compute the distances between consecutive bifurcation points and list your results below. Now compute the ratios of the consecutive distances (the n-th distance divided by the (n+1)-th distance). What do you notice about these ratios? (If you make the correct observation, then the number in your observation is what is referred to as Feigenbaum's constant.) Place your computations and comments below:

```
>
```

Make a conjecture (on the basis of the evidence at hand) about the next distance ratio, and then about the distance to the next bifurcation point. Place your conjectures below:

```
> BiData := [???];
> BiDataDiff := [seq(BiData[n+1]-BiData[n], n=1..7)];
> BiDataRat := [seq(BiDataDiff[n]/BiDataDiff[???], n=1..6)];
```

¤ 6. Additional investigations (Optional).

Find a 3-cycle by searching the bifurcation diagram for a likely location. Confirm (or refine) your estimate by constructing a web diagram that actually shows the 3-cycle. Enter your computations and comments below.

```
>
```

Starting with a three cycle, attempt to find a six cycle, a twelve cycle, and a twenty-four cycle. Moreover, attempt to find good estimates for the bifurcation points that separate such cycles, and check to see if the bifurcation points in this case have the same relationship with Feigenbaum's constant as we found with the previous set of bifurcations. Enter your computations and comments below.

```
>
```

You have seen that bifurcation points emerge faster and faster as lambda increases, and now you have a good idea--from Feigenbaum's constant--of how much faster. Calculate the smallest value of lambda by which you think a bifurcation of every order will have appeared, i.e., by which a 2^n-cycle for every positive integer n will have appeared. (Hint: To answer this question in a rigorous way you will need to determine the limit of

1 + x + x^2 + x^3 + x^4 + x^n

for |x|<1 as n increases without bound. This is known as a geometric series.) Place your computations and comments below:

```
>
```

□ LABORATORY RESULTS Summary Sheet.

Complete as much of this table as you can.

Bifurcation number	Number of branches	Estimated lambda value	Distance from previous lambda	Ratio to next dist.
1	2	____________		
2	4	____________	_______________	__________
3	8	____________	_______________	__________
4	16	____________	_______________	__________
5	32	____________	_______________	__________
6	64	____________	_______________	__________
7	128	____________	_______________	__________
8	256	____________	_______________	__________
9	512	____________	_______________	__________
10	1024	____________	_______________	

Conjecture: The next distance ratio is approximately ____________.

Hence the distance to the __-th bifurcation is approximately _____.

This means the __-th bifurcation occurs at approximately lambda = _______________.

The three cycle. Complete as much of this table as you can. (Optional)

Bifurcation number	Number of branches	Estimated lambda value	Distance from previous lambda	Ratio to next dist.
1	3	____________		
2	6	____________	_______________	__________
3	12	____________	_______________	__________
4	24	____________	_______________	__________
5	48	____________	_______________	__________
6	96	____________	_______________	

Does the same distance ratio seem to hold true for bifurcations starting with the three-cycle as for the bifurcations starting with a two-cycle? ____________.

LAB 16: Area From Sums

□ NAMES:

□ PURPOSE.
Our goal is to understand how to approximate areas under curves by sums of rectangles or trapezoids. Understanding this approximation is basic to understanding the definite integral.

□ PREPARATION.
Read the beginning of Chapter 8, pages 459 to 469, in Calculus: Modeling and Application.

□ LABORATORY REPORT INSTRUCTIONS.
Note: Follow your instructor's instructions if they differ from those given here.

For this laboratory you are not required to submit a formal report. Rather you are asked to submit a copy of this Worksheet with the questions answered in the indicated locations. Only one submission will be made. Write in complete sentences and be sure the sentences connect together in coherent ways.

□ Maple Procedures. Please activate the following line.

```
> with(lab16);
> # If an Error is the output of the above line, see your lab instructor.
```

□ PRELIMINARIES.

¤ Summations.

In this laboratory experiment we will deal with large sums (unfortunately not money). For this reason we begin by examining the way Maple handles calculation of sums. Here's a summary to aid your understanding.

"Intuitive" meaning: a1 + a2 + a3 + ... + a(n-1) + an

Maple command: sum(a(k), k=1..n)

Use the Maple command, sum, to compute each of the following summations. Then check your answers by activating the command line which follows each summation.

1 + 2 + 3 + ... + 49 + 50

```
> sum(???);
> Answer(1);
```

5^2 + 6^2 + ... + 19^2 + 20^2

```
> sum(???);
> Answer(2);
```

(1.5)^4 + (1.5)^5 + ... + (1.5)^12 + (1.5)^13

```
> sum(???);
> Answer(3);
```

The Distributive Rule for addition and multiplication,

c*a1 + c*a2 + ... + c*an-1 + c*an = c*(a1 + a2 + ... + an-1 + an)

yields the following rule for sum:

sum(c*a(k), k=1..n) = ____________________________ . Check your answer by activating the following line.

```
> Answer(4);
```

The Commutative and Associative Rules for addition allow the following rearrangement:

(a1 + b1)+(a2 + b2)+...+(an-1 + bn-1)+(an + bn) = (a1 + a2 + ... + an-1 + an)+(b1 + b2 + ... + bn-1 + bn)

This formula yields the following rule for sum:

sum(a(k)+b(k), k=1..n) = ____________________________ . Check your answer by activating the following line.

```
> Answer(5);
```

The rules given above can be used to simplify statements involving sums. Well-chosen simplifications can have a positive effect on the speed of computation.

□ PROJECT.

¤ 1. Choosing the Region Under Consideration.

We will begin our investigation of area by choosing a region to analyze in the xy plane. We wish to consider a region as shown below: bounded below by the x-axis, bounded on the sides by horizontal lines at x=a and x=b, and bounded above by the graph of a function y=f(x).

```
> Region(1+2*x-x^3, x=0..1.5);
```

You need to define such a function f(x) (no, it does not have to be the function shown in the picture) and the bounds a and b. Enter f(x), a, and b below.

```
> f := x -> ???; a := ???;  b := ???;
> Region(f(x),x=a..b);
```

Since we need to have the function f(x) above the x-axis for x between a and b, then the function must be non-negative on the interval (a, b). To determine if your function is always positive on the interval we have provided a convenient checking routine: PositivityTester. Enter the following command.

```
> PositivityTester(f(x), x=a..b);
```

If your function did not check out as being non-negative, then change your choice of f(x) and/or a and b. Continue using PositivityTester until you obtain a function f(x) which is non-negative on (a,b).

When you finally have settled on a function f(x) and an interval (a,b), plot them using the command Region. You obtain information on Region in the usual way, i.e., by Entering ?Region. Enter the following special plotting command.

```
> Region(f(x), x=a..b);
```

Make a crude estimate as to how much area there is in the region shown in your plot. Explain briefly what method you used to obtain your estimate. Enter your estimate as a Maple statement defining the quantity areaEstimate. Complete the following command, then Enter the result.

```
> areaEstimate := ???;
```

Answer: The method used to estimate the area was...???

¤ 2. Left-Hand sums--A Direct Computation.

You will need f(x), a, b, and areaEstimate from the previous section. To check that they have been properly defined, Enter the following command.

```
> f(x);  a;  b;
```

We now consider more formalized methods for estimating the area chosen in Section 1. The first method we use--left-hand sums--is the basis of the definition for the "definite integral."

We divide the base of the region (i.e., the interval [a,b]) into n equal pieces, where n is a positive integer of your choice. Enter your choice for n as a Maple statement. Enter your statement below.

• Maple Tip. Your subsequent pictures will not look good unless n <= 10.

```
> n := ???;
```

We divide the interval [a,b] into n equal pieces and build rectangles over each piece using the left-hand endpoint. We have a special command--Riemann--to generate a picture for this collection of rectangles with your function f(x). Enter the following command.

```
> Riemann(f(x), x = a..b, n,left);
```

How well does the sum of the areas of the rectangles (shown in light green) approximate the area under the curve y=f(x)? Briefly justify your answer. Place your answer below.
Answer:

A primary goal of this section is to compute the sum of the areas of the rectangles shown above.

We know that the area of each rectangle is the product of its width times its height. The width--Deltax--is easy to compute. Enter a formula for Deltax in terms of the symbols a, b, and n. This formula should return the numerical value of Deltax for your example. Enter your formula for Deltax below.

```
> Deltax := ???;        # Use the symbols a, b and n in your formula.
```

Using Deltax and a it is now possible to obtain a formula for each of the subdivision points x(0), x(1), x(2), ..., x(n-1), x(n). Such a formula will be necessary when computing the heights of our rectangles. First complete the following table:

x(0) := a
x(1) := a + ???
x(2) := ???
x(3) := ???

You should see the pattern in these four entries, and hence be able to write down a general formula: x(k) = ???

This general formula allows us to define all of the x(k) in one seq command. It helps to use a capital x here, X, to distinguish it from the variable x in the expression f(x). Complete and Enter the following command.

```
> X := k -> evalf(???);    # Use the symbols a, k, and Deltax in your formula.
> seq(X(k), k= 0..n);
```

• Maple Tip. It is convenient to use evalf in your definition of X or to make sure that each X(k) has a decimal point in it to obtain the numbers in decimal form. As you can see, the definition of X above incorporates this.

We are now ready to determine the height of each rectangle. Let's proceed with the rectangles one at a time, starting with the first. For convenience we generate a picture of just the first rectangle by using LHSPlotter again. Enter the following Riemann command.

```
> Riemann(f(x), x=a..b, [n, 1],left);
```

Determine the area, area(1), of this first rectangle--your formula should be in terms of the symbols x(0), f, and Deltax. Enter your formula as a Maple command. (Hint: What is the height of the rectangle in terms of X(0) and f ?) Complete and Enter the following statements.

```
> area(1) := evalf(???); # Use the symbols X(0), f, and Deltax in your formula.
```

We now perform similar computations for the second rectangle. First generate a picture for the second rectangle by using LHSPlotter. Enter the following LHSPlotter command.

```
> Riemann(f(x), x=a..b, [n, 2],left);
```

Determine the area, area(2), of this second rectangle--your formulas should be in terms of the symbols x(1), f, and Deltax. Enter your formulas as Maple commands. Complete and Enter the following statements.

```
> area(2) := ???;
> # Use the symbols X(1), f, and Deltax in your formula.
```

Hopefully you're ready to compute the area of a "typical" rectangle in our left-hand sum. The following command will generate a picture of a "typical" k-th rectangle. Enter the following command.

```
> k := ???:
> Riemann(f(x), x=a..b, [n, k],left);
```

Determine the area, area(k), of this k-th rectangle. Your formula should be in terms of the symbols x(...), f, and Deltax. Record your answer here (but don't Enter it):

area(k) := ???;   # Use the symbols X(), f, and Deltax in your formula.

The formula for area allows us to calculate all the desired areas in one seq command. Complete and Enter the following seq command.

```
> area := k -> ???:   # Use the symbols X( ), f, and Deltax in your formula.
> seq(area(k), k= ???);
```

Re-plot the original LHSPlotter command that showed all n of the rectangles, then briefly compare the area numbers you just obtained with the areas of the rectangles shown in the picture. Do your numbers look correct? Enter the following command.

```
> Riemann(f(x), x=a..b, n,left);
```

Finally we can compute the sum of the areas of our rectangles. This is just an application of the sum command (as considered in Preliminaries) to the quantities area(k). Complete and Enter the following command.

```
> k := 'k':
> totalSum := sum ???
```

Compare this value with areaEstimate, your estimate of the area under the curve from the end of the previous section. Are they close? Do you wish to alter your estimate in light of the computations in this section? If so, Enter your new value for areaEstimate below. Enter your commands and answers below.

```
>
```

Summary--left-hand sums.

For convenience in the next section (where we will automate our procedures), fill in the following summary of the major steps we went through to compute our left-hand sums. Fill in the blanks below

Start with a function f(x) on an interval a <= x <= b and a positive integer n. Then...

1. Compute the length Deltax of the n equal subintervals of [a,b].
   Deltax := ???; Ans: ???

2. Determine the division points X(i).
   X := k -> ???: Ans: ???

3. Determine the areas of the sub-rectangles, area(k).
   area := k -> ???: Ans: ???

4. Compute the total area of all the rectangles combined, totalSum.
   totalSum := ???; Ans: ???

¤ 3. Left-Hand Sums--An Automated Procedure.

You will need f(x), a, b, and areaEstimate from the previous sections.

In the previous section you fixed an integer n and computed the left-hand sum with this number of rectangles. However, if done correctly, your commands should be equally valid for any value of n. To check if this is true,

• gather together the four input statements from the summary at the end of the previous section,
• place them into one region, and
• Enter the commands with a variety of different n values.

What happens to totalSum as n increases? Enter your commands and answers below.

```
> n := 6:  k := 'k':
> Deltax := ???:                        # These are the four commands that
> X := k -> ???:                        # will be referred to later.
> area := k -> ???:
> totalSum :=  evalf(sum(area(k), k=???));
```

Answer:

Our next goal is to show you how to write a Maple routine LHSums that will compute left-hand sums for any (reasonable) combination of f(x), a, b, and n.
First, we need to discuss some aspects of Maple programming.

Maple Lesson.

• Maple Tip. Inside a procedure it is dangerous to define functions with the -> command. It may work, as in the following procedure, where the function g is defined without passing any information from the main procedure to it.

Activate the following region then verify that the procedure H works by calculating H(1), H(2), ... in the next region.

```
> H := proc (x)
>   local  g, k;
>   g := k -> 2*k;
>   g(x);
> end:
> ???
```

Next, let's pass some information into the definition of g.

```
> H1 := proc (x)
>   local  g, k, d;
>   d := 2;
>   g := k -> d*k;     # The constant  d  needs to be passed into  A.  It is not!
>   g(x);
> end:
> ???
```

You can see that now there is trouble, the value assigned to d was not passed into the function g. This can be fixed by defining g in another way.

• Maple Tip. Within procedures, you can pass information into functions that are defined with the unapply( ) command.

Think of the unapply( ) command as follows: Given a function f, we use the terminology of "applying" the function f to the variable k to obtain the expression f(k). So it would be reasonable to "unapply" the expression f(k) to obtain the function f. Thus, f := unapply(f(k), k). The k after the comma is telling Maple what symbol represents the variable.

```
> H2 := proc (x)
>  local g, k, d;
>  d := 2;
>  g := unapply(d*k ,k);
>  g(x);
> end;
> ???
```

Back to computing Left-Hand Sums.

We are now ready to go back and use the four commands referred to before the Maple Lesson.

Place your four commands from above into the "shell" for LHSums that is given below (i.e., replace the ????????????? seen below with your commands as modified according to the Maple Lesson above).

```
> LHSums := proc(f,a,b,n)
> local Deltax, X, area, totalSum, k;
>
> ??????????
> end;
```

Here are some checks on the accuracy of your new routine. Enter the following statements--the output of your routine LHSums should match the answers given below. Enter the following commands and examine the results.

```
> LHSums(f,  0, 1.5,4);
```

The answer should be 2.264648437 .

```
> LHSums(x->x^2, -1, 2, 40);
```

The answer you should obtain is 2.890312500 .

```
> LHSums(x -> sin(x)^2, 1, 2, 40);
```

The answer you should obtain is 0.9149538463 .

Now that LHSums is available to us, let's combine it with the plotting routine Riemann in the following useful way. Using the function and interval that you chose, we will start the n values at n=2, but then increase them by doubling at each stage, i.e., 2, 4, 8, 16, 32, and 64. Let's see what happens to both the numbers generated by LHSums and the pictures generated by Riemann. Enter the following set of commands.

```
> frame := NULL:
> for k from 1 to 6 do
>   frame := frame,Riemann(f(x), x=a..b,2^k,left);
> od:
> plots[display]([frame],insequence=true);
```

We have a Maple command "movie" that automates the above animation.

```
> movie(f(x),x=a..b,6,left);
```

We now show you how to list values for LHSums.

```
> printf(` n          LHSums for n\n`);
> for k from 1 to 6 do
>   printf(`%2d          %f\n`, 2^k,LHSums(f,a, b, 2^k));
> od:
```

You have now seen the six pictures produced by the movie command and the corresponding values giving the areas under the approximations. What do you see in the pictures? And what do you see in the numbers generated by LHSums? Do the behavior of the pictures and the behavior of the numbers seem consistent with each other? Place your answer below.

Answer:

Try a large value of n in LHSums, say n=400. What do we obtain now? Enter your commands below.

```
> ???
```

Answer:

Do you wish to make any additional alterations to areaEstimate, your approximation to the area under the curve? Enter your statements below.

```
> ???
```

¤ 4. Approximations by Trapezoids.

You will need f(x), a, b, areaEstimate and LHSums from the previous sections.

There is a relatively minor alteration that can be made in the definition of LHSums which will produce a surprisingly better approximation to the area under the graph of a function. To motivate this change, execute the following command.

```
> Riemann(f(x),x=a..b,3,trap,left);
```

As the above picture should suggest, we intend to replace rectangular approximations of area with trapezoidal approximations. Recall your formula for the area of a left hand sum rectangle. Place your answer below (Don't enter this).

area(k) := ??? (the area of the k-th left-hand sum rectangle)

Alter this formula to give the area of the first trapezoid. Refer to the plot to see what has to be changed. (Hint: Make use of the average height of the trapezoid.) Place your answer below (Don't enter this).

area(k) := ??? (the area of the k-th trapezoid)

Use this area formula to construct a new Maple routine TrapSums that will approximate the area under graphs of functions with trapezoids instead of rectangles. (Hint: simply take the commands used in LHSums and change one line of code!!) Enter your commands and answers below.

• Maple Tip. Use Copy, Paste, and an alteration of the corresponding portions of LHSums definition is the easiest way to construct TrapSums.

```
> TrapSums := proc(f,a,b,n)
> local Deltax,X,area,totalSum,k;
>
> ??????????
>
> end;
```

Here are some checks on the accuracy of your new routine. Enter the following statements--the output of your routine TrapSums should match the answers given below. Enter the following commands and examine the results.

```
> TrapSums(x ->x^2,  -1, 2, 40);
```

The answer should be 3.002812500

```
> TrapSums(x -> sin(x)^2, 1,  2, 40);
```

The answer should be 0.91643820...

Now that TrapSums is available, let's combine it with the plotting routine Riemann in the following useful way. Using your function f(x) and interval (a,b), we will start with n = 2, then increase it by doubling at each stage, i.e., 2, 4, 8, 16, 32, and 64 Let's see what happens to both the numbers generated by TrapSums and the pictures generated by Riemann. Enter the following set of commands.

```
> movie(f(x),x=a..b,6,trap); #This animates the  6  pictures generated by Riemann.
```

Let's now look at the corresponding values computed by LHSums and TrapSums.

```
> printf(` n    LHSums for n    TrapSums for n\n`); for k from 1 to 6 do
>  printf(`%2d      %f          %f\n`,2^k,LHSums(f,a,b,2^k),TrapSums(f,a,b,2^k));
> od:
```

You have seen the six pictures and the corresponding values of the areas of the approximation. What do you see in the pictures? And what do you see in the numbers generated by TrapSums? Do the behavior of the pictures and the behavior of the numbers seem consistent with each other? Place your answer below.
Answer:

Which method--left-hand sums or trapezoidal approximations--do you prefer for approximating area? Justify your choice by reference to both (1) accuracy of the approximations for a fixed n, and (2) speed of convergence to the "exact" area as n increases. Place your answers below.
Answer:

Apply your new routine TrapSums to the example you extensively studied in the previous sections, say with n=5, 40, and 400. On the basis of these numbers do you wish to make any changes in your areaEstimate? Enter your commands and answers below.

```
>

> areaEstimate := ???;
```

¤ 5. The Bell Curve. (Optional--Check with your Instructor.)

You will need LHSum and TrapSums from the previous sections. To check that they have been properly defined, Enter the following command.

```
> LHSums(x->x^2, -1, 2, 6);  TrapSums(x->x^2, -1, 2,6);
```

If any of the items in the list failed to produce a reasonable value, then go back and reenter the commands which defined them.

*****

We will use the two approximation schemes developed above to approximate the area of the region that is bounded above by the bell curve,

```
> plot(sqrt(1/(2*Pi))*exp(-(x^2/2)),x=-5..5);
```

below by the x-axis, and on the sides by the lines x = xmin = 0 and x = xmax = 3. Complete and Enter the following function definition command.

• Maple Tip. Maple knows the base of the natural exponential function as E, not "e". The exponential function (with base E) can also be denoted by exp(...).

```
> g := x -> ???;     # The bell curve function.
> xmin := ???;
> xmax := ???;
```

First we generate plots of the two area approximation procedures by the use of Riemann. Simply choose a value for m, the number of subintervals you want in the interval [0,3], and the commands below will show both the left-hand sum approximation and the trapezoid approximation to the area. The numerical values of these approximations are then printed out under the plots. Does it appear that one method is preferable to the other? Justify your choice. Complete and Enter the commands below, then state the method you prefer.

• Maple Tip. You will probably want to expand the picture generated below.

```
> m := ??? :
> Riemann(g(x),x=xmin..xmax,m,left,trap);
> printf(`When m = %d\n`,m);
> printf(`left-hand sums equal %f\n`,LHSums(g, xmin,xmax, m));
> printf(`trapezoid sums equal %f\n`, TrapSums(g, xmin,xmax, m));
>
```

Answer:

Use either LHSums or TrapSums to obtain as accurate an approximation as you can to the area under the bell curve from x = 0 to x = 3. Indicate how many digits of your final answer you believe to be accurate (and give some justification for your claim). Enter your computations and explanations below.

```
>
```

¤ 6. Right-Hand Sums. (Optional--Check with your Instructor.)

We now consider the area-estimating method of right-hand sums. As you will see, this requires just a slight alteration of the definition of left-hand sums. In fact, the only item that gets changed is the height used for each of the rectangles. Let's compare a right-hand sum rectangle with a left-hand sum rectangle by means of the following pictures. Enter the following set of commands.

You may need to change the order of the plotting to get a good view. This can be accomplished by changing the order of left and right in the Riemann command.
Try to figure out why this is the case.

```
> Riemann(f(x),x=a..b,[4,2],right,left);
```

Recall your formula for the area of the left hand sum rectangles from the previous sections. Complete the following formula.

area(k) := ??? (a left hand sum rectangle)

Now alter this formula to give the area of the right hand sum rectangles. Refer to the two plots shown above to see what has to be changed. Complete the following formula.

area(k) = ??? (a right hand sum rectangle)

Use this area formula to construct a new Maple routine RHSums that will approximate the area under graphs of functions with right-hand sums rather than left-hand sums. (Hint: simply take the commands used in LHSums and change one line of code!!) Enter your commands and answers below.

• Maple Tip. Use Copy, Paste, and an alteration of the corresponding portions of LHSums is the easiest way to proceed here.

```
> RHSums := proc(f,a,b,n)
> local Deltax, X, area, totalSum, k;
>
> ?????????????
>
> end;
```

Here are some additional checks on the accuracy of your new routine. Enter the following statements--the output of your routine RHSums should match the answers given below. Enter the following commands and examine the results.

```
> RHSums(x -> x^2, -1, 2, 40);
```

The answer should be 3.115312500

```
> RHSums(x -> sin(x)^2, 1, 2, 40);
```

The answer should be 0.917922556

Now that RHSums is available to us, let's combine it with the plotting routine Riemann in the following useful way. Using the function and interval that you chose, we will start the n values at n=2, but then increase them by doubling at each stage, i.e., 2, 4, 8, 16, 32, and 64 Let's see what happens to both the numbers generated by RHSums and the pictures generated by RHSPlotter. Enter the following set of commands.

```
> movie(f(x),x=a..b,6,right);
> printf(` n    RHSums for n\n`);
```

```
> for k from 1 to 6 do
>     printf(`%2d    %f\n`,2^k, RHSums(f, a, b, 2^k));
> od;
```

You should now have seen the six pictures and the corresponding values of the approximating areas. What do you see in the pictures? And what do you see in the numbers generated by RHSums? Do the behavior of the pictures and the behavior of the numbers seem consistent with each other? Place your answers below.

Answer:

As you can probably see, the right-hand sum approximation to an area is in general no better than the left-hand sum approximation. From that point-of-view you might well wonder why we have bothered to introduce the right-hand sum method. Here is one answer: for certain types of areas, the left-hand sums and the right-hand sums will define a range of numbers which we know contains the exact value of the area. In this way we can obtain error bounds on our approximations which we know are absolute, i.e., the exact value of the area cannot fall outside of this range. A concrete example of this situation now follows.

We consider again the bell curve g(x) over the interval [0,3], as discussed in the previous section. First we generate plots of the left-hand and right-hand sums by the use of Riemann. Simply choose a value of m, the number of subintervals you want for the interval [0,3], and the commands below will show both the left-hand and right-hand sum approximations to the area. The numerical values of these approximations are then printed out under the plots. Think about the relationship between these two approximations and the actual area A. Complete and Enter the commands below.

• Maple Tip. You will probably want to expand the picture generated below.

```
> m:=10:
> Riemann(g(x),x=xmin..xmax,m,left_right);
> printf(`When m = %d,\n`,m);
> printf(`    Right-hand sums equal %f and \n`,RHSums(x -> g(x), xmin, xmax, m));
> printf(`     Left-hand sums equal %f\n`,LHSums(x -> g(x), xmin, xmax, m));
```

Notice that the bell curve g(x) is decreasing over the whole range of our x interval. What special relationship does that guarantee will hold true between any left-hand sum, any right-hand sum, and the exact area A under the graph of y=g(x)? Place your answers below.

Answer:

Use the fact just stated to determine the area of our region to at least two decimal places, i.e., to an error of no more than .005. Can you obtain an even better set of bounds on the actual area A? How does your answer to this question compare with your findings at the end of the previous section? Enter your computations and final answer below. □

LAB 17A: Functions Defined by Integration

□ NAMES:

□ PURPOSE

Many extremely important functions are naturally defined by the integration of known functions. Sometimes these turn out to be functions we already know; sometimes they are new. In this lab we examine several examples and study the question of what can be said about derivatives of functions defined by integration. Then we consider the result of performing the two operations in the other order: first differentiation, then integration. Underlying both of these investigations are the First and Second Fundamental Theorems of Calculus.

□ PREPARATION

Read sections 8.2 through 8.4, pgs 483 - 498, in Calculus: Modeling and Application.

□ LABORATORY REPORT--General Instructions.

Note: Follow your instructor's instructions if they differ from those given here.

You are to write a self-contained, well-organized exposition and analysis of the issues raised in this laboratory. In particular, it should incorporate the answers to all the important questions raised during the lab session.

Two submissions will be made for this report: an initial draft, and a final revision. Comments made on the draft will help you prepare the revision, and only the revision will receive a grade. Moreover, the grade will be based solely on the written report--the instructor will look at other parts of the worksheet only in unusual situations. You should therefore consider the entries you make in the Project section to be notes to help when later writing your report.

More specific instructions on the content of the report will be given at the end of the worksheet.

□ Maple Procedures. Activate the following line.

```
> with(lab17a);
> # If an Error is the output of the above line, see your lab instructor.
```

□ PRELIMINARIES

¤ Symmetric Difference Quotients and Numerical Derivatives.

Suppose s is a quantity which depends on a variable t, and we have measurements for s at the t-values t1 < t2 < t3. How could we numerically estimate the derivative of s at t2? We will need to make such estimates later in this lab.

Our first thought, since we are working with data at a number of discrete t values, would be to approximate the derivative at t=t2 by the difference quotient

$$Ds/Dt = (s(t3)-s(t2))/(t3 - t2)$$

We claim, however, that a better estimate to the derivative is generally obtained by using what we shall call the symmetric difference quotient

(Ds/Dt)sym = (s(t3)-s(t1))/(t3 - t1)

As you can see, to approximate the derivative at t2 via symmetric difference quotient (SDQ) we use the function points to either side of t2, i.e., (t1, s(t1)) and (t3, s(t3)).

To convince you that, in most cases, the symmetric difference quotient is indeed the better approximation to the derivative, we present some graphical evidence. We have defined a Maple command for this lab called sdqPlotter. Use the ?sdqPlotter command to find out about it.

Enter the following commands.

```
> ?sdqPlotter
> func := t -> t^3;
> t2 := 1.0;
> run := .5;
> sdqPlotter(func, t2, run);
```

Explanation: The function func(t) = t^3 is plotted for the interval 0.5 <= t <= 1.5 . Three points are shown as large dots, corresponding to t1=0.5, t2=1.0, and t3=1.5 . Delete the incorrect choices for each question concerning the center point t2.

The difference quotient is the slope of the
    (red, blue, or magenta) line.
The symmetric difference quotient is the slope of the
    (red, blue, or magenta) line.
The derivative is the slope of the
    (red, blue, or magenta) line.
The derivative is best approximated by which quotient?
    ("regular" or symmetric)

Now this is just one example--perhaps the outcome is different for most other functions! Well, one of the powerful features of Maple is that you can generate as many examples as you like! The command sdqPlotter is designed to generate LOTS of examples--all you need to do is make the desired changes. You can change the function func(t), the t-value t2, and the distance "run" between t2 and t1 (and between t2 and t3). So that you don't destroy the good example generated above, a copy of the input statements are given below for your alteration. Try some of your own examples, leaving the "best" one you are able to generate as an additional illustration of the relationships between the derivative, the difference quotient, and the symmetric difference quotient. Enter a few examples of your own below.

Suggestion: It is instructive (and amusing) to try and find aberrant situations in which the symmetric difference quotient is worse than the (ordinary) difference quotient as an approximation to the derivative. Such examples are interesting, however, only if run is small.

```
> func := t -> ???; t2 := ???; run := ???;
> sdqPlotter(func, t2, run);
```

For the convenient use of symmetric difference quotients in the numerical approximation of derivatives, we have written a special command, nD ("numerical derivative"). This command will approximate the derivative of any function by a symmetric difference quotient. Use the ?nD command for more information. Like the D command in Maple, the nD command, written for this lab, is designed to have as an argument a function written with the arrow command, and it returns a function. Enter the following commands.

```
> f := x -> x^3:
> nD(f);
> nD(f)(x);
> simplify(nD(f)(x));
```

Is nD(f) a good approximation to the derivative D(f) of f? Why? Place your answer below.
Answer:

Here are examples using a version of ND constructed for evaluation at a specific x=x0 value. Enter the following examples of ND.

```
> nD(x ->x^5)( 2);
> g := x -> sin(x);
> nD(g)(0);
```

Are these values good approximations for the specified derivatives? Place your answers below.
Answer:

□ PROJECT.

¤ 1. Choosing the Function Under Consideration.

We will begin our investigation by choosing a function y=f(x) to analyze. We wish our function to define a region as shown below: bounded below by the x-axis, bounded on the sides by horizontal lines at x=a and x=b, and bounded above by the graph of y=f(x).

```
> graph(1);
```

You need to define such a function f(x) (no, it does not have to be the function shown in the picture) and the bounds a and b. Enter f(x), a, and b below.

```
> f := x-> ???; a := ???: b:= ???:
```

Since we need to have the function f(x) above the x-axis for x between a and b, then the function must be non-negative on the interval (a, b). In addition (for reasons to become clear later), we need to have the derivative of f(x) exist everywhere on (a,b). It is also convenient to have f(a) not equal zero (otherwise you might be misled on a conjecture later in the lab). To determine if your function satisfies these conditions on (a,b), we provide three convenient checking routines: PositivityTester, DerivativeTester, and InitialValueTester. Enter the following commands.

```
> ?PositivityTester
> ?DerivativeTester

> PositivityTester(f(x), x=a..b);
> DerivativeTester(f(x), x=a..b);
```

We define the initialValueTester here. Activate the following region and then test it on your function.

```
> initialValueTester := proc(f, a, b)
>   if evalf(f(a))=0 then lprint(`f(x) is zero at x=a.  Fix this.`);
>   else   lprint(`f(x) is NOT zero at x=a.  Good!`); fi; end:
> initialValueTester(f, a, b);
```

If your function did not check out as being non-negative, or your derivative did not exist at some of the sample points, or f(a) is zero, then change your choice of f(x) and/or a and b. Continue using PositivityTester, DerivativeTester, and InitialValueTester until you obtain an acceptable function f(x) and interval (a,b).

Note: Our "testing commands" are not fool-proof, i.e., there may be times your function f(x) is given a clean bill-of-health when in fact it has a problem. Hopefully this will rarely occur. However, do not abandon all responsibility for checking f(x) to the commands--you should analyze the function as well.

When you finally have settled on a function f(x) and an interval (a,b), plot them using the command Region. You obtain information on Region in the usual way, i.e., by Entering ?Region. Enter the following special plotting command.

• Maple Warning! For the Region command we use the calling sequence
Region(f(x), x=xmin..xmax), where
f(x) stands for an expression in x.

```
> fPlot :=Region(f(x), x=a..b):";
```

¤ 2. Functions Defined by Integrals.

You will need f, a, and b from the previous section. To check that they have been properly defined, Enter the following command.

```
> f(x); a; b;
```

If any of the items in the list failed to produce the correct value, then go back to Section 1 and reenter the commands which defined them.

* * * * *

Our intention is to define a new function F by a definite integral of the original function f. Enter the next command to give our definition of F.

```
> F :=x-> int(f(t),t=a..x);
```

(Notice the two different independent parameters t and x. The original function f is a function of t; however, when we integrate f from the t values t=a to t=x, the result is a quantity which varies with x, i.e., a function F of the variable x. Said another way, before we integrate, the independent variable is t, but after we integrate the independent variable is x.)

Using Maple's numerical integration command int, compute F(x) for specific numerical values of x between a and b.

```
>
```

In order to better understand the geometric significance of F(x), vary the values of x between a and b, using the special graphics command given below. Enter the following command for a variety of x values, $a < x < b$:

```
> x := ???:
> plotFpicture(f,a,b,x,20);
```

From these plots, and from the defining equation for F, describe geometrically the meaning of the number F(x) for a fixed value of x. Place your answer below.

Answer:

Plot the graph of y=F(x) for the values $a <= x <= b$, and explain (using the geometric interpretation you obtained above) why the graph takes the form that it does. Enter your plot below, along with your explanation.

```
> x := 'x':
> ???
```

Explanation:

• Maple Tip. If your Plot command generates an error message then go back and reenter the commands which defined them.

¤ 3. Derivatives of Integrals.

You will need f(x), a, b, and fPlot from the previous sections. To Check that they have been properly defined, Enter the following command.

```
> x := 'x': f(x); a; b; fPlot;
```

If any of the items in the list failed to produce the correct value, then go back and reenter the commands which defined them. (fPlot should produce a graphic.)

* * * * *

We have seen in the previous section how to define a function F from the definite integral of a given function f:

```
> F := x -> int(f(t),t=a..x);
```

$$F := x \rightarrow \int_a^x f(t)\,dt$$

Determining the behavior of the derivative of F(x) is one of the two major objectives of this laboratory project.

To compute the derivative of F(x) we recall that the derivative is a limit of difference quotients:

F'(x) = limit((F(x+Deltax)-F(x))/Deltax, Deltax = 0);

We will compute the derivative F'(x) by obtaining an estimate for the numerator term F(x+Deltax)-F(x) of the difference quotient. To obtain this estimate we will make use of the geometric interpretation for F(x) examined in the previous section. This geometric interpretation is usefully illustrated in the graphics which follow. To generate these plots,

(1) set Deltax equal to 0.05,
(2) pick a value of x so that x and x+Deltax both lie between a and b, and
(3) place these values into the commands given below.

Define the x and Deltax values below, then Enter the commands.

• Maple Tip. Notice that the pictures you are about to generate "zoom in" on and magnify the size of the yellow strip between x to x+Deltax, making it easier to see. As a result of this magnification we see only a portion of the region over the interval a <= t <= b. In general the full purple region extends further to the left, while the full white region extends further to the right. Only the yellow region is seen in its entirety.

```
> x := ???:
> Deltax := ???:
> plotFstrip(f, a, b, x, Deltax, 50);
```

Questions: Express the areas of the following regions in terms of F(x) and F(x+Deltax):

(1) the purple region,
(2) the purple + yellow region, and
(3) the yellow region.

Record your answers below by completing the following statements.

The purple region = ????
The purple + yellow region = ????
The yellow region = ????

Is there a simple estimate for the area of the yellow region in terms of the original function f? If so, the answer should be given in terms of the symbols f, x, and Deltax--don't use specific numbers for these quantities. (Hint: Does the yellow region resemble any simple geometric shape?) Place your estimate below.

Simple estimate for the area of the yellow region = ????

Execute plotFstrip two or three more times, using successively smaller values of deltax. What happens to the shape of the yellow strip as Deltax decreases? (Remember that the magnification increases as deltax decreases.) Continue to look for a simple estimate for the area of the yellow region in terms of the original function f. Does your estimate become more or less accurate as deltax decreases? Enter your plots and explanations below.

```
> x := ???:
> Deltax := ???:
> plotFstrip(f, a, b, x, Deltax, 50);
```

Simple estimate for the area of the yellow region = ????

Explanation:

You have obtained two expressions for the area of the yellow region, one exact and one an approximation. List your two expressions below. Place your two expressions below.

The yellow region = _____???_____
(an expression using F, x, and deltax)

Give a good approximation for the difference quotient for F, at least when Deltax is small. (Hint: equate the two expressions you have for the area of the yellow region.) Place your approximation formula below.

(F(x+Deltax)-F(x))/Deltax ~ ________________

Based on this approximation, conjecture a formula for the derivative F'(x) of F at x. Explain how your conjectured formula follows from the approximation in the previous question. Hint: use the definition of the derivative of F(x) as the limiting value of difference quotients,

F'(x) = limit((F(x+Deltax)-F(x))/Deltax, Deltax = 0)

Place your conjecture in the space below, followed by your explanation.

¤ "Conjecture A" : F'(x) =

Explanation:

Checking Conjecture A.

We can get a good estimate for the derivative of F(x) at any point by using symmetric difference quotients, as discussed in the Preliminaries. This is easily carried out by using the numerical differentiation command nD. Then we can generate a plot of the (approximate) derivative of F (denoted as FDer) and see if it supports "Conjecture A". Enter the following commands defining FDer.

```
> x := 'x':
> F(x);
> FDer := nD(F);
```

Plot the derivative function y=FDer(x) on the interval a <= x <= b. Call this graphic FDerPlot. (Be patient--this plot can sometimes take a bit of time.) Enter your plot command below.

```
> FDerPlot := plot(???):";
```

We now compare the plot of the derivative of F (FDerPlot) with the elaborate plot of the original function f that we made in Section 1 (fPlot) by placing them both in the same xy-coordinate plane. What do you conclude from this comparison? Enter the following command, followed by your conclusions.

```
> plots[display]([fPlot,FDerPlot]);
```

Conclusion:

Does this plot support your "Conjecture A" for the derivative of F? Explain. Place your observations below.

On the basis of your results, complete the following expanded version of "Conjecture A".

Taking a function f, integrating it from a to x, and then differentiating the result with respect to x will yield ....

¤ 4. Integrals of Derivatives.

You will need f(x), a, b, F(x), and fPlot from the previous sections. To check that they have been properly defined, Enter the following command.

```
> x := 'x':
> f(x); a; b; fPlot; F(x);
```

If any of the items in the list failed to produce the correct value, then go back and reenter the commands which defined them. (fPlot should produce a graphic.)

* * * * *

We again start with our function f(t) on the interval a <= t <= b. However, instead of integrating to get F(x) and then differentiating, this time we will first differentiate f(t) and then integrate the result.

Since we wish to continue working numerically,

• first use the nD command to obtain fDer(t) as an approximate derivative of f(t) (this is how FDer(x) was obtained from F(x) in the previous section),

• second, use the int command to obtain H(x) as an approximate definite integral of fDer(t) from t=a to t=x (this is how F(x) was obtained from f(t) in Section 2).

Complete and Enter the following Maple code to define fDer(t) and H(x).

```
> fDer :=nD ???
> H := x -> int ???
```

• Determining the behavior of the function H(x) is the second major objective of this laboratory project.

Plot the resulting function y=H(x) over a <= x <= b. (Be patient--this plot may take a little time.) Enter your plotting command below.

```
> Hplot := ???
```

How does the graph of H relate to the graph of f? In order to most easily answer that question, we show both the graph of H (Hplot) and the graph of f (fPlot) on the same xy-coordinate axes. Enter the following command, then state your conclusions.

```
> plots[display]([Hplot, fPlot]);
```

Conclusion:

As a result of your observations, conjecture a simple formula for the function H(x). This is the second of the two major conjectures in this laboratory project. Place your conjecture in the space below.

¤ "Conjecture B" H(x) =

Complete the following expanded version of "Conjecture B".

Taking a function f, differentiating it, and then integrating the result from a to x will yield ....

¤ 5. All of the Above for the Bell Curve. (Optional--Ask your instructor.)

Repeat Sections 2, 3, and 4 for the function f(t):=exp(-0.5*t^2) on the interval a=0 to b=3. Do the results of these computations continue to support the conjectures you made earlier? In particular, you should do the following (and comment on the results):

Basic Theory:

Plot y=f(t) to have the graph for reference.
Produce at least one plot showing a region whose area is F(x).
Plot the graph of the function F(x).
Show a plot illustrating whether or not your "simple approximation"
for F(x+Dx)-F(x) is still valid for f(t)=Exp(-.5*t^2).

Enter your commands and notes for the "Basic Theory" below.

¤ "Conjecture A":

Plot the derivative of F, and use it to determine if "Conjecture A" is still valid for f(t)=Exp(-.5*t^2).

Enter your commands and notes for "Conjecture A" below.

¤ Conjecture B":

Plot the function H, and use it to determine if "Conjecture B" is still valid for f(t)=Exp(-.5*t^2).

Enter your commands and notes for "Conjecture B" below.

¤ 6. All of the Above for the Sine Curve. (Optional--Ask your instructor.)

Repeat Section 5 for the function f(t)=sin(t) on the interval a = 0 to b = Pi/2. Do the results of these computations continue to support the conjectures you made earlier? Enter your commands and notes below.

¤ 7. The Fundamental Theorems of Calculus.

Explain the relationship of your conjectures with the First and Second Fundamental Theorems of Calculus. In particular, can you use one of the Fundamental Theorems to strengthen one of your two conjectures? Place your answers and comments below.

□ Laboratory Report--Specific Instructions.

Write a self-contained, well-organized exposition of the issues raised in this lab. In particular, your report should not be a simple step-by-step restatement of the project--that shows little recognition nor analysis of the central ideas of the lab. Instead, as with any good expository paper, you should begin with a clear understanding of the principal ideas you wish to communicate, and then decide on a structure for your report that will best convey these ideas. You can then discuss specific portions of the lab project in the appropriate parts of the report.

An example of a "generic" organization that can be used for a laboratory report is to divide your essay into four sections:

(1) Purpose and goals.
(2) Procedures.
(3) Results.
(4) Analysis.

Remember to keep your audience in mind: you should assume your reader is comfortable with the concepts of calculus developed thus far in the course, but is not familiar with the particular material of this laboratory.

Explain the conjectures you made, how and why you changed them, and their final status. Discuss any observed differences in behavior among the functions you considered (if you worked with more than one function f(t)). The instructions in Section 5 of the Project can be used as an initial outline for your report.

You should discuss similarities and differences between Conjectures A & B and the following two statements: (a + b) - b = a and (a - b) + b = a, where addition is associated with integration and subtraction is associated with differentiation.

Finally, explain the relationship of your conjectures with the Fundamental Theorems of Calculus. In particular, can you use the Fundamental Theorem to strengthen one of your conjectures?

Place your report below.

□

LAB 17B: The Fundamental Theorem of Calculus

□ NAMES:

□ PURPOSE
In this laboratory we derive the Fundamental Theorem of Calculus from Euler's Method. We also study the inverse relationship between differentiation and definite integration.

□ PREPARATION
Read sections 8.2 through 8.4, pages 483-498, in Calculus: Modeling and Application.

□ LABORATORY REPORT--General Instructions.
Note: Follow your instructor's instructions if they differ from those given here.

You are to write a self-contained, well-organized exposition and analysis of the issues raised in this laboratory. In particular, it should incorporate the answers to all the important questions raised during the lab session.

Two submissions will be made for this report: an initial draft, and a final revision. Comments made on the draft will help you prepare the revision, and only the revision will receive a grade. Moreover, the grade will be based solely on the written report--the instructor will look at other parts of the worksheet only in unusual situations. You should therefore consider the entries you make in the Project section to be notes to help when later writing your report.

More specific instructions on the content of the report will be given at the end of the worksheet.

□ Maple Packages.

```
> with(lab17b);
> # If an Error is the output of the above line, see your lab instructor.
```

□ PRELIMINARIES

¤ 1. The One-Step Euler's Approximation.

During the Project we will give a justification for the Fundamental Theorem of Calculus based on the following observation.

Suppose we wish to approximate F(b)-F(a) for a function F(x). Our method will be derived from rewriting F(b)-F(a) in the following way:

$$F(b)-F(a) = ((F(b)-F(a))/(b-a)) (b-a)$$

The quotient in the large parentheses is merely the average rate of change of F(x) over the interval from x=a to x=b. However, if b-a is small, then this average rate of change will not differ very much from the instantaneous rate of change of F(x) at x=a. This instantaneous rate of change is merely the derivative of F(x) at x=a, i.e., F'(a). Putting these observations together yields:

$$F(b)-F(a) = F'(a)\,(b-a), \quad \text{when } b-a \text{ is small}$$

We will call this the One-Step Euler's Approximation for F(b)-F(a). Geometrically, the Euler's Approximation F'(a)(b-a) is the length of the opposite side of a right triangle formed by an adjacent side equal to b-a and a hypotenuse of slope F'(a). This hypotenuse is merely the tangent line to y=F(x) at x=a.

Let's look at some examples. Enter the following command.

```
> EulerSteps(sin, 0, .8, 1);
```

The blue curve is the graph of F(x)=sin x, and the thick blue vertical line has length equal to

$$F(b)-F(a) = \sin .8 - \sin 0 = \sin .8.$$

The thick red vertical line has length equal to the One Step Euler's Approximation,

$$F'(a)\,(b-a) = (\cos 0)\,(.8-0) = 0.8.$$

Examine what happens to Euler's Approximation as the difference b-a is increased and decreased. Generate further examples by completing and Entering the following command.

```
> a := ???: b := ???:
> EulerSteps(sin, a, b, 1);
```

How good is the Euler's Approximation when b-a is small? What about when b-a is large? Place your answers below.

Answer:

¤ 2. The Definite Integral.

Since it will appear in the Project, we recall the definition of the definite integral. First, we use the symbol as shown by activating the following region to stand for the definite integral of f from a to b.

```
> a := 'a': b := 'b':
> Int(f(t), t=a..b);
```

The definition.
Fix an integer n and partition the interval a <= t <= b into n equal pieces of length Deltat=(b-a)/n.
The partitioning points will be given by

t[k] = a + k*Deltat, for k = 0, 1, ..., n

Given this notation, the definite integral is defined to be the number to which the following summation converges as n goes to infinity.
Activate the following line to see the appropriate summation symbolism.

```
> Sum(f(t[k])*Delta*t, k=0..n);
```

• Maple Tip. In the command line written above, the Delta*t was written in that way as a means of getting the correct looking computer output. Actually, it should be Deltat and is equal to (b-a)/n .

When this limit arises, you need to recognize it as the definite integral.

¤ 3. Symmetric Difference Quotients and Numerical Derivatives.

Suppose s is a quantity which depends on a variable t, and we have measurements for s at the t-values t1 < t2 < t3. How could we numerically estimate the derivative of s at t2? We will need to make such estimates later in this lab.

The most obvious approach, since we are working with data at a number of discrete t values, would be to approximate the derivative at t=t2 by the difference quotient

Ds/Dt = (s(t3)-s(t2))/(t3 - t2)

We claim, however, that a better estimate to the derivative is generally obtained by using what we shall call the symmetric difference quotient

(Ds/Dt)sym = (s(t3)-s(t1))/(t3 - t1)

As you can see, to approximate the derivative at t2 via symmetric difference quotient (SDQ) we use the function points to either side of t2, i.e., (t1, s(t1)) and (t3, s(t3)).

To convince you that, in most cases, the symmetric difference quotient is indeed the better approximation to the derivative, we present some graphical evidence. Enter the following commands.

```
> func := t -> t^3: t2 := 1.0: run := .5:
> sdqPlotter(func, t2, run);
```

Explanation: The function func(t)= t3 is plotted for the interval 0.5 <= t <=1.5 . Three points are shown as large dots, corresponding to t1=0.5, t2=1.0, and t3=1.5 . Delete the incorrect choices for each question concerning the center point t2.

The difference quotient is the slope of the
(red, blue, or dashed magenta) line.
The symmetric difference quotient is the slope of the
(red, blue, or dashed magenta) line.
The derivative is the slope of the
(red, blue, or dashed magenta) line.
The derivative is best approximated by which quotient?
("regular" or symmetric)

Now this is just one example--perhaps the outcome is different for most other functions! Well, one of the powerful features of Maple is that you can generate as many examples as you like! The command sdqPlotter is designed to generate LOTS of examples--all you need to do is make the desired changes. You can change the function func(t), the t-value t2, and the distance "run" between t2 and t1 (and between t2 and t3). So that you don't destroy the good example generated above, a copy of the input statements are given below for your alteration. Try some of your own examples, leaving the "best" one you are able to generate as an additional illustration of the relationships between the derivative, the difference quotient, and the symmetric difference quotient. Enter a few examples of your own below.

Suggestion: It is instructive (and amusing) to try and find aberrant situations in which the symmetric difference quotient is worse than the (ordinary) difference quotient as an approximation to the derivative. Such examples are interesting, however, only if run is small.

```
> func := t -> ???;
> t2 = ???; run = ???;
> sdqPlotter(func, t2, run);
```

For the convenient use of symmetric difference quotients in the numerical approximation of derivatives, we have written a special command, nD ("numerical derivative"). This command will approximate the derivative of any function by a symmetric difference quotient; the default value of the run is .0001, although any run can be specified by use of the option sdqDelta. (More information on nD can be obtained with the usual "?" command.) Enter the following examples of nD.

```
> f := x -> x^3:
> nD(f);
> nD(f)(x);
> simplify(nD(f)(x));
```

Is nD(f) a good approximation to the derivative D(f) ? Why? Place your answer below.

Answer:

Here are examples using a version of nD constructed for evaluations at a specific x = x0 value. Enter the following examples of nD.

```
> nD(x ->x^5)( 2);
> g := x -> sin(x);
> nD(g)(0);
```

Are these values good approximations for the specified derivatives? Place your answers below.
Answer:

□ PROJECT

¤ 1. The n-Step Euler's Approximation.

We will begin our investigation by choosing a function y=F(t) to analyze. We also need to pick an interval a <= t <= b over which to work. Choose a and b so that b-a is relatively large. Enter F, a, and b below.

```
> F :=???; a := ???: b := ???:
```

When you have settled on a function F and an interval (a,b), plot them using the command region. (You can obtain information on region in the usual way, i.e., by Entering ?Region.) Enter the following special plotting command.

• Maple Warning! For the Region command we use the calling sequence Region(f(x), x=xmin..xmax), where f(x) stands for an expression in x.

```
> FPlot:=Region(F(x), x=a.. b):";
```

Let's attempt to estimate F(b)-F(a) by the use of the One-Step Euler's Approximation. (Does your plot give you reasons to think this will be a good approximation or a bad one?) Enter the following command.

```
> EulerSteps(F, a, b, 1);
```

Hopefully you did not obtain a very good approximation (if you did, then go back and increase the size of b-a). The remedy is to take more than one "Euler step"--let's start by taking two Euler steps. Enter the following command.

```
> EulerSteps(F, a, b, 2);
```

Did you obtain much improvement in your approximation for F(b)-F(a) in going to 2 steps? Place your answer below.

Answer:

All we did was to divide the interval a <= t <= b into two pieces and perform an Euler's Approximation on each piece. In equations this is what we have (we will let f(t) denote the derivative of F(t)):

Let Deltat = (b-a)/2 and t0=a, t1=a+Deltat, t2=a+2*Deltat=b. Then

```
F(b)-F(a) = (F(t1)-F(t0)) + (F(t2)-F(t1))
      ~=    f(t0)*Deltat  +  f(t1)*Deltat
```

We can keep playing this game, i.e., increase the number of Euler's steps from n=2 to n=3. Enter the following command.

```
> n := 3 :
> EulerSteps(F, a, b, n);
```

Did you obtain much improvement in the approximation for F(b)-F(a) in going from 2 to 3 steps? Place your answer below.

Answer:

All we did was to divide the interval a <= t <= b into three pieces and perform an Euler's Approximation on each piece. In equations this is what we have:

Let Deltat = (b-a)/3 and t0=a, t1=a+Deltat, t2=a+2*Deltat, t3=a+3*Deltat=b. Then

```
F(b)-F(a) = (F(t1)-F(t0)) + (F(t2)-F(t1)) + (F(t3)-F(t2))
      ~=    f(t0)*Deltat   +  f(t1)*Deltat   +  f(t2)*Deltat
```

Increase the number of Euler's steps from n=3 to n=8. Then to n=16. Then to n=32. Is the approximation continuing to improve? Complete and Enter the following command for several values of n.

```
> n := ???;
> EulerSteps(F, a, b, n);
```

Answer:

Define a new function and try the Euler's Approximation again. Complete and Enter the following commands.

```
> a := ???: b := ???:
> G := x -> ???;

> GPlot := Region(G(x), x=a..b):";
> n := 7:
> for i from 1 to n do
>   frame[i] := EulerSteps(G, a,b,2^i): printf(`.`);od:
> plots[display]([seq(frame[i], i=1..n)], insequence=true);
```

The limiting value of the n-Step Euler Approximation for F(b)-F(a) as n increases can be seen symbolically by activating the following command line.

```
> n := 'n': f := 'f':
> Limit (Sum(f(t[k])*Delta*t, k=0..n) ,n=infinity);
```

$$\lim_{n \to \infty} \sum_{k=0}^{n} f(t_k)\,\Delta\,t$$

State the major formula that this gives for F(b)-F(a), and justify your answer. What is the name of this result? Place your answer below.

Answer:

¤ 2. Integrals of Derivatives.

What happens if we start with a function F, take its derivative, and then compute the definite integral of the derivative from t=a to t=x? Let's look at some examples. Complete and Enter the following commands.

Define a function F(t):

```
> F := t -> ???; a := ???: b := ???:
```

Plot the function F:

```
> FPlot := Region(F(x), x=a..b):";
```

Differentiate the function F(t), then integrate the result from t=a to t=x to get H(x):

```
> f := D(F); H := x -> evalf(int(f(t), t=a..x));
```

Plot H(x) (be patient--this plot may take a little time), then compare it with F(x)

```
> HPlot := plot(H, a..b,color=green):";

> plots[display]({FPlot, HPlot});
```

How does the graph of H relate to the graph of F? Place your answer below.

Answer:

Go back to the start of this section, change the function F(t), and re-execute the commands which follow. Does the relationship between the graphs of H and F which you observed above still seem to be true? Place your answer below.

Answer:

As a result of your observations, conjecture a simple formula for the function H(x). Place your conjecture in the space below.

"Conjecture" H(x) =

Complete the following expanded version of the "Conjecture".

Taking a function F, differentiating it, and then integrating the result from a to x will yield ....

Relate your conjecture to the final result which you obtained at the end of Section 1. In particular, can you improve your conjecture by using the Section 1 result? Place your answer below.

Answer:

¤ 3. Derivatives of Integrals.

What happens if we start with a function f(t), take its definite integral from t=a to t=b, and then compute the derivative with respect to x? Let's look at some examples. Complete and Enter the following commands.

Define a function f(t):

```
> f := t -> ???; a := ???: b := ???:
```

Plot the function f(t):

```
> fPlot := Region(f(x), x=a..b):";
```

Integrate the function f(t) from t=a to t=x, then differentiate the result with respect to x:

```
> F := x -> int(f(t), t=a..x);
> h := nD(F);
```

Plot h(x) , then compare it with f(x).

```
> hPlot := plot(h, a..b,color=green):";
> plots[display]([fPlot, hPlot]);
```

How does the graph of h relate to the graph of f? Place your answer below.

Answer:

Go back to the start of this section, change the function f(t), and re-execute the commands which follow. Does the relationship between the graphs of h and f which you observed above still seem to be true? Place your answer below.

Answer:

As a result of your observations, conjecture a simple formula for the function h(x). Place your conjecture in the box below.

"Conjecture" h(x) =

Complete the following expanded version of the "Conjecture".

Taking a function f, integrating it from a to x, and then differentiating the result with respect to x will yield ....

How does this conjecture relate to the conjecture made at the end of Section 2? Place your answer below.

Answer:

Remark: The conjecture you should have observed in this Section is essentially a result known as the Second Fundamental Theorem of Calculus.

□ LABORATORY REPORT--Specific Instructions.

Write a self-contained, well-organized exposition of the issues raised in this lab. In particular, your report should not be a simple step-by-step restatement of the project--that shows little recognition nor analysis of the central ideas of the lab. Instead, as with any good expository paper, you should begin with a clear understanding of the principal ideas you wish to communicate, and then decide on a structure for your report that will best convey these ideas. You can then discuss specific portions of the lab project in the appropriate parts of the report. An example of a "generic" organization that can be used for a laboratory report is to divide your essay into four sections:

(1) Purpose and goals.
(2) Procedures.
(3) Results.
(4) Analysis.

Remember to keep your audience in mind: you should assume your reader is comfortable with the concepts of calculus developed thus far in the course, but is not familiar with the particular material of this laboratory.

Explain the conjectures you made, how and why you changed them, and their final status. Discuss any observed differences in behavior among the functions you considered.

You should discuss similarities and differences between your Conjectures in Sections 2 and 3 and the following two statements: (a + b) - b = a and (a - b) + b = a, where addition is associated with integration and subtraction is associated with differentiation.

Finally, explain as much as you can about the relationship of your conjectures with the Fundamental Theorems of Calculus. In particular, can you use the Fundamental Theorem to strengthen one of your conjectures?

Place your report below.

□

LAB 18: Inverse Functions

□ NAMES:

□ PURPOSE
In this lab we examine the concept of inverse pairs of functions. We consider a number of different examples, including the exponential, natural logarithm, sine, and tangent functions. It will be seen that certain of the corresponding inverse functions are most conveniently expressed in terms of definite integrals. Conversely, these inverse functions (especially the arcsine and arctangent) furnish us with new antiderivatives, and hence important integration formulas.

□ PREPARATION
Review Section 1.8, pages 49-58, entitled "Inverse Functions" in Chapter 1 of Calculus: Modeling and Application. Before coming to the lab, read over the Preliminaries and answer the questions found in that section--the answers will prepare you for a computation to be performed during the project.

□ LABORATORY REPORT INSTRUCTIONS
For this laboratory you are not required to submit a formal report. Rather you are asked to submit:

¤ An electronic copy of this Worksheet with the questions answered in the indicated locations. Only one submission will be made. Write in complete sentences and be sure the sentences connect together in coherent ways.

¤ A filled-in paper copy of the Laboratory Results Summary Sheet.

□ MAPLE PROCEDURES
Please activate the following command.

```
> with(lab18);
> # If an Error is the output of the above line, see your lab instructor.
```

Lab 18 procedures includes two new commands---InversePlot, and InverseFunctionCheck --- which will be described as needed later in the lab.

□ PRELIMINARIES

¤ Some Trigonometry Facts.

For what values of q in -Pi <= q <= Pi is cos(q) positive? Write your answer down on paper, then check it by activating the following line.

```
> Answer(1);
```

For what values of q in -Pi <= q <= Pi do we have cos(q) = sqrt(cos^2(q))? Why? Write your answer down on paper, then check it by activating the following line.

> **Answer(2);**

For what values of q in -Pi <= q <= Pi do we have cos(q) = sqrt(1-sin^2(q))? Why? Write your answer down on paper, then check it by activating the following line.

> **Answer(3);**

¤ Inverse Functions.

We recall some basic facts concerning inverse pairs of functions. You should further review the material on inverse functions on pages 52-58 in Chapter 1 of Calculus: Modeling and Application. Write your answers down on paper, then check it by activating the command line below.

¤ The functions f and g are inverse functions for each other if
f(a)=b is true if and only if ..... is true.

> **Answer(4);**

¤ The functions f and g are inverse functions for each other if
g(f(x))= .... is true for all x in the domain of f, and
f(g(x))= .... is true for all x in the domain of g.

> **Answer(5);**

¤ The functions f and g are inverse functions for each other if
(a,b) is on the graph of y=f(x) if and only if ..... is on the graph of y=g(x).

> **Answer(6);**

This last condition implies that the graphs of inverse functions f and g are the mirror images of each other across the line y=x, as shown in the following picture.

> **Graphic(1);**

This graph also helps clarify the relationship between the domain and range of a function f and the domain and range of its inverse g: they get interchanged. Answer the following and check it.

¤ If the functions f and g are inverse functions for each other, then
the domain of f is the ..... of g, and the range of f is the ..... of g.

> **Answer(7);**

□ PROJECT.

¤ 1. When does an inverse function exist?

The command InversePlot(f, a..b)

(1) plots the graph of y=f(x) over the interval [a,b] in red, and

(2) plots the reflection of this graph across the line y=x in a dashed curve. We can use this plot to determine if f(x), when restricted to [a,b], has an inverse, as follows:

If (1) f(x) is defined for every value of x in [a,b], and

(2) the reflected (dashed) curve is actually the graph of a function (i.e., passes the "vertical line test"), then f(x), when restricted to (a,b), has an inverse function.

Use InversePlot to determine which of the following functions have inverses. Record results below and in Table A of the Laboratory Results Summary Sheet.

• Maple Tip. The following list of command lines has been set up for easy use.

(1) Enter a command line defining f, a, and b, and then

(2) Enter the InversePlot command found at the bottom of the list. This will produce the desired plot.

```
> f := x -> x^2; a:=0; b:=2;              # Has inverse? Yes, No.
> f := x -> x^2; a:=-2; b:=2;             # Has inverse? Yes, No.
> f := x -> sqrt(4-x^2); a:=0; b:=1;      # Has inverse? Yes, No.
> f := x -> sqrt(4-x^2); a:=-1; b:=1;     # Has inverse? Yes, No.
> f := x -> log(x); a:=1; b:=50;          # Has inverse? Yes, No.
> f := x -> log(x); a:=-1; b:=2;          # Has inverse? Yes, No.
>                              # Careful on this one. Is log(x) defined for all a<x<b?
> f := x -> sin(x); a:=-Pi; b:=2*Pi;      # Has inverse? Yes, No.
> f := x -> sin(x); a:=0; b:=Pi/2;        # Has inverse? Yes, No.
> f := x -> tan(x); a:=0; b:=1.57;        # Has inverse? Yes, No.
> f := x -> tan(x); a:=0; b:=2*Pi;        # Has inverse? Yes, No.

> InversePlot( f, a .. b );
```

Important! Does the existence of an inverse for a function f(x) depend on the interval [a,b] to which we have restricted f(x)? Said in other words, might f(x) have an inverse on one interval [a1,b1] but not have an inverse on another interval [a2,b2]? Cite examples to support your claim. Place your answer below.

¤ 2. Determining the largest interval on which an inverse function exists.

For suitable restrictions of each of the functions considered in Section 1, i.e.,

x^2, sqrt(4-x^2), log(x), sin(x), and tan(x),

we want to find a formula for the inverse function. In this section we take an important step in this direction by determining two intervals for each function f(x) on our list:

- the largest interval [a,b] containing zero such that f(x), when restricted to [a,b], will have an inverse g(x), and

- the interval [c,d] on which this inverse g(x) is defined.

We accomplish this by using the command InversePlot(f, a .. b) to determine the largest interval [a,b] such that the potential inverse curve (dashed) is the graph of a function. Then, from the plot so obtained, we read off the interval [c,d] on which the inverse function is defined.

As an example, the following plot was generated by using [a,b]=[1,4] as the domain of the original red function. The range of this function is seen to be [c,d]=[1,8]. Hence, for the inverse dashed function, the domain and range are interchanged, i.e., the domain of the inverse dashed function is [c,d]=[1,8], and the range is [a,b]=[1,4].

```
> Graphic(2);
```

For each of the following functions, determine [a,b] and [c,d]. Record your results below and in Table B of the Laboratory Results Summary Sheet.

• Maple Tip.

The following list of command lines has been set up for easy use. For each,

(1) type in guesses for a and b and Enter the line (which defines f, a, and b), then

(2) Enter the InversePlot command found at the bottom of the list.

On the basis of the plot so produced you can either

(1) alter the values of a and b and run the plot again, or

(2) conclude that you found the best a and b values, and hence read off the c and d values.

(Note: All statements after a #) are "invisible" to Maple. Hence you do not have to fill in values for c and d before executing the command lines.)

• Maple Tip. If you are tempted to use -infinity for a or infinity for b, don't, since InversePlot will respond with an error message. Instead use finite values of "modest" size for a and b. Often this is sufficient to understand the full behavior of the inverse function.

Interval Determinations.

```
> f := x -> x^2; a:=???: b:=???:             #b=???, c=???, d=???
> f := x -> sqrt(4-x^2); a:=???: b:=???:     # c=???;  d=???
> f := x -> log(x); a:=???: b:=???:          #a=???, b=???, c=???, d=???
> f := x -> sin(x); a:=???: b:=???:          # c=???;  d=???
> f := x -> tan(x); a:=???: b:=???:          # c=???;  d=???

> InversePlot( f, a .. b );
```

¤ 3. Identifying formulas for inverse functions.

Formulas exist for the inverses of all the functions considered in the previous section (although some of the formulas are more complicated than you might expect). Momentarily we will give a list which contains among them all five of these inverses--your task is to pick the inverses out of the list. Again, the command InversePlot is used to make the necessary identifications. In the form

InversePlot(f, a..b, g)

the command plots f in red, the potential inverse as a dashed curve, and g in blue. You compare the dashed and blue curves; if they match, then g is the inverse function.

Before giving our list, we need to introduce a few new functions, AS(x) and AT(x). They are defined via the definite integral (ah...finally a connection to integration), in the same way that we defined functions via the definite integral in the previous lab:

Activate the following region to see how they look in standard math language.

```
> AS(x) = Int(1/sqrt(1-t^2),t = 0 ..x);
> AT(x) = Int(1/(1+t^2),t=0..x);
```

$$AS(x) = \int_0^x \frac{1}{\sqrt{1-t^2}}\,dt$$

$$AT(x) = \int_0^x \frac{1}{1+t^2}\,dt$$

As Maple functions we define them via the numerical integration routine evalf. Enter the following function definition statements.

```
> AS := x -> evalf( Int( 1/sqrt(1-t^2), t = 0 .. x) );  #  -1 < x < 1;
> AT := x -> evalf( Int( 1/(1+t^2), t = 0 .. x) );      # -infinity < x < infinity;
```

The following is the list which contains the inverses of the five functions listed at the end of the last section. Identify (and test) the five inverse functions.

1/x^2,	sqrt(x),
1/sqrt(4-x^2),	sqrt(4-x^2),
1/log(x),	exp(x),
1/sin(x) (i.e., csc(x)),	1/tan(x) (i.e., cot(x)),
AS(x),	AT(x)

For your convenience we again list the five functions under study. In each case you need to fill in the values for a and b that you determined in the previous section. Moreover, in each case you should guess at the appropriate inverse function g(x), and type it into the designated spot. Check your guesses with InversePlot. Record your results below and in Table B of the Laboratory Results Summary Sheet.

• Maple Tip. The following list of command lines has been set up for easy use. For each,

(1) type in your values for a and b (as determined in the previous section) and your guess at the inverse function g, then Enter the line and

(2) Enter the InversePlot command found at the bottom of the list.

On the basis of the plot so produced you can either

(1) alter the choice of g and run the plot again, or

(2) conclude that g is indeed the inverse function for the function on the interval [a,b].

The graph of g is blue, so what you are looking for is the blue line and the dashed line coinciding. (Note: The values of c and d are not used here, but will be important in the next section.)

Inverse function determination.

```
> f := x -> x^2:  a:=???:  b:=???:              g:= x -> ???;
> f := x -> sqrt(4-x^2): a:=???: b:=???:        g:= x -> ???;
> f := x -> log(x): a:=???: b:=???:             g:= x -> ???;
> f := x -> sin(x): a:=???: b:=???:             g:= x -> ???;
> f := x -> tan(x): a:=???: b:=???:             g:= x -> ???;
> InversePlot( f,  a .. b, g );
```

¤ 4. Checking the inverse function pairs: undoing each other.

From Section 3 you now have determined--at least graphically--five inverse function pairs. Now we check them against one of the characterizations of inverse function pairs that we looked at in the Preliminaries section:

• The functions f and g are inverse functions for each other if

g(f(x))= .... is true for all x in the domain of f, and

f(g(x))= .... is true for all x in the domain of g.

Let's make this rule specific to our current situation. Let f(x) denote any one of our original five functions--the domain of f is the interval [a,b]. Let g(x) denote the (probable) inverse function for f(x), as determined in the previous sections--the domain of g is the interval [c,d]. With this notation, our characterization of inverse functions thus becomes.

• The functions f and g are inverse functions for each other if

g(f(x))= .... is true for all x in the interval ...., and

f(g(x))= .... is true for all x in the interval .....

It will be easy to check this condition for the five inverse function pairs because we have a graphical command InverseFunctionCheck which plots g(f(x)) and f(g(x)) over the designated intervals. To do this most conveniently, complete the following set of commands for our inverse function pairs and then apply InverseFunctionCheck. Be sure to specify accurately the two intervals [a,b] and [c,d] for each function pair! Record results below and in Table B of the Laboratory Results Summary Sheet.

• Maple Tip. The following list of command lines has been set up for easy use. For each,
(1) complete the line, then Enter it and
(2) Use the InverseFunctionCheck command found at the bottom of the list. Should your plots not come out as desired (??), then you need to check over your previous work.

```
> f := x -> x^2: a:=???: b:=???:            g:= x -> ???;   c:=???:   d:=???:
> f := x -> sqrt(4-x^2): a:=???: b:=???:   g:= x -> ???;   c:=???:   d:=???:
> f := x -> log(x): a:=???: b:=???:         g:= x -> ???;   c:=???:   d:=???:
> f := x -> sin(x): a:=???: b:=???:         g:= x -> ???;   c:=???:   d:=???:
> f := x -> tan(x): a:=???: b:=???:         g:= x -> ???;   c:=???:   d:=???:
>
> InverseFunctionCheck(f, a..b, g, c..d);
```

• Maple Tips.
1) For the intervals [a, b] ,[c, d] where you would like to use and infinity for either b or d, use finite values and determine in which cases the actual values should be infinity.

Assuming that for each function, f, you made the correct choices for g, a, b, c, and d, then all of the plots you have just generated should look basically "the same." Briefly describe the nature of this plot and explain why this plot had to arise if f on [a,b] and g on [c,d] are indeed inverse functions for each other. Place your discussion below.

¤ 5. The arcsin Function.

The derivative of AS(x). From the definition of AS(x) as a definite integral you should be able to quickly compute the derivative of AS(x). Do so, and be sure to say how you got the answer. Place your answer below.

The arcsin function. Maple has a built-in function, arcsin(x), that is related to the functions sin(x) and AS(x). Experiment with specific values of the three functions by evaluating all three at various values of x. Also construct a plot of arcsin(x). Conjecture a relationship between arcsin(x), sin(x), and AS(x). Place your computations and answers below.

The derivative of arcsin(x). Use the derivative you found previously for AS(x), and your conjectured relationship between AS(x) and arcsin(x), to conjecture a derivative formula for arcsin(x). Then check your conjectured derivative formula using Maple's ability to compute derivatives, i.e., with the diff(h(x), x) command. Does your conjectured formula match what Maple's diff command produces? Place your computations and answers below.

```
> Darcsin := diff(???,x);
> DAS := ???;
> plot(Darcsin, x=???);
> plot(DAS, x=???);
```

The definition of arcsin function. From your work thus far, you should see that the arcsin is the .... for the restricted sine function--in fact, that is how the arcsin function is defined! Hence you can obtain a simple formula for sin(arcsin(x)), and a specification of the x-values for which the formula is valid. Place your formula below.

sin(arcsin(x))= ?? for x values in the interval ?? < x < ??.

The derivative of arcsin(x)--a theoretical derivation. Now a paper-and-pencil exercise (you may wish to do this outside of the lab period). The formula for the derivative of the arcsin function can be obtained directly from your just-stated formula for sin(arcsin(x)) by applying the Chain Rule and the results of the Preliminaries section. Do this! Place your computations in the Laboratory Results Summary Sheet.

Application to definite integration. Using the results of this lab project, evaluate each of the following definite integrals in two different ways, first using AS(x), then using arcsin(x):

```
A := Int(1/sqrt(1-t^2),t=0..0.75);
B := Int(1/sqrt(1-t^2),t=-0.3..0.4);
```

Place your computations below.

¤ 6. The arctan Function.

The derivative of AT(x). From the definition of AT(x) as a definite integral you should be able to quickly compute the derivative of AT(x). Do so, and be sure to say how you got the answer. Place your answer below.

The arctan function. Maple has a built-in function, arctan(x), that is related to the functions tan(x) and AT(x). Experiment with specific values of the three functions by evaluating all three at various values of x. Also construct a plot of arctan(x). Conjecture a relationship between arctan(x), tan(x), and AT(x). Place your computations and answers below.

The derivative of arctan(x). Use the derivative you found previously for AT(x), and your conjectured relationship between AT(x) and arctan(x), to conjecture a derivative formula for arctan(x). Then check your conjectured derivative formula using Maple's ability to compute derivatives, i.e., with the diff(h(x), x) command. Does your conjectured formula match what Maple's diff command produces? Place your computations and answers below.

The definition of arctan function. From your work thus far, you should see that the arctan is the .... for the restricted tangent function--in fact, that is how the arctan function is defined! Hence you can obtain a simple formula for tan(arctan(x)), and a specification of the x-values for which the formula is valid. Place your formula below.

tan(arctan(x))= ?? for x values in the interval ?? < x < ??.

The derivative of arctan(x)--a theoretical derivation. Now a paper-and-pencil exercise (you may wish to do this outside of the lab period). The formula for the derivative of the arctan function can be obtained directly from your just-stated formula for tan(arctan(x)) by applying the Chain Rule. Do this! Place your computations in the Laboratory Results Summary Sheet.

Application to definite integration. Using the results of this lab project, evaluate each of the following definite integrals in two different ways, first using AT(x), then using arctan(x).

```
A := Int(1/(1+t^2),t=0..0.75);
B := Int(1/(1+t^2),t=-0.3..0.4);
```

Place your computations below.

□ LABORATORY RESULTS SUMMARY SHEET.

¤ Table A. Which of the following functions have inverses on the specified intervals?

Function	Interval	Has an Inverse[Y/N]?	If not, why not?
x^2	[0,2]		
x^2	[-2,2]		
sqrt(4-x^2)	[0,1]		
sqrt(4-x^2)	[-1,1]		
log(x)	[1,infinity)		
log(x)	[-1,2]		
sin(x)	[-Pi, 2*Pi]		
sin(x)	[0, Pi/2]		
tan(x)	[0,Pi/4]		
tan(x)	[0,2Pi]		

¤ Table B. Determining the inverses for five common functions.

Function	Largest allowable Interval [a,b]	Largest allowable interval [c,d] for the inverse	Formula for inverse function	Checked formula? (Y/N)
x^2				
sqrt(4-x^2)				
log(x)				
sin(x)				
tan(x)				

¤ Derivative Computations.

Place your computations for the derivatives of arcsin(x) and arctan(x) below. □

LAB 19: Symbolic Computation

□ NAMES:

□ PURPOSE

We have used Maple primarily as a super-powerful, programmable, graphics calculator and text processor. Now we will explore its ability to perform symbolic computation, e.g., algebraic manipulations, symbolic differentiation, symbolic integration, and calculation of solutions of algebraic and differential equations. This lab is merely an introduction to Maple's vast powers in this arena.

□ PREPARATION

You should work through the first few chapters of Andre Heck, Introduction to Maple. We will repeat some of this material, but the focus will be on additional topics, especially (1) the algebraic manipulation of polynomials and rational functions, and (2) symbolic differentiation and symbolic integration.

□ MAPLE PROCEDURES

```
> with(lab19);
> # If an Error is the output of the above line, see your lab instructor.
```

□ LABORATORY REPORT INSTRUCTIONS

Note: Follow your instructor's instructions if they differ from those given here.

For this project you are not required to submit a formal report. Rather you are asked to submit a copy of this worksheet at the end of the lab period with the questions answered in the indicated locations. This laboratory will be ungraded--or, more precisely, it will be graded "Pass/Fail"--the only requirement is that you submit the completed Worksheet.

□ PRELIMINARIES

We remind you of certain basic Maple facts that are generally useful.

¤ The symbol " refers to the output of the last executed Maple command. Likewise "" refers to the output of the command Before the last executed command.

¤ Always end commands with either a ; or a : . The difference between the two is in output. If a command is ended with a ; , then the output will be displayed to the screen. On the other hand if a : is used, then no output is displayed. The " command works in either case.

¤ When assigning variables, the output will be usually be displayed in a different form than if the value had been entered directly. To get the output you normally get use :"; at the end of your assignment. i.e. A := plot(x^2,x=0..5):"; This will assign A the plot data, and also puts a plot on the screen.

¤ Blocks of commands can be combined and separated using the Format options: Join (F4) and Split (F3) groups. The F4 key removes the separator above the cursor and the F3 key adds a separator above the cursor. This allows the combining of several commands into one group for sequential execution.

¤ Use the ? command to get help on any Maple command. i.e. ?solve

□ PROJECT.

¤ 1. Operations with Polynomials.

• Factorization.

In this section we examine the basic operations that Maple can perform on polynomials. Enter the following command to recall the general form of a polynomial function.

```
> poly();
```

$$a_0 + a_1 x + a_2 x^2 + a_3 x^3 + \ldots + a_n x^n$$

where the a[k] are all constants.

Begin by Entering the following fifth degree polynomial:

```
> P := 3*x^5-18*x^4-7*x^3+42*x^2-40*x+240;
```

Now factor this polynomial by the use of the (need we say it?) factor command:

```
> factor(P);
```

Unfortunately factor will not break quadratic terms (i.e., ax^2+bx+c) down to linear terms (i.e., ax+b) if they involve square roots or complex numbers. Here's how to work around this problem and achieve a total factorization of this polynomial. Start by solving for the roots. To do this, Enter the following solve command:

```
> S := solve(P=0, x);
```

• Explanation of the solve command.

Input to solve. The single equality P=0 is not an assignment statement, rather it is a logical statement which, for each value of x, produces either the value true or false.

Example: x^2=4 is a logical statement which is true for either x:=2 or x:=-2, but is false for all other x-values.

The solve command solve(P=0,x) merely determines those values of x which make the statement P=0 take the value true. Such values of x are, of course, the roots of the polynomial in question.

Output from solve. We have obtained five values in a sequence from solve, one for each root.

One last feature of the output from the solve command deserves mention: the meaning of the "I" which appears in the values for two of the roots. This stands for the square root of -1, i.e., what in high school you probably denoted by i. As this demonstrates, Maple is able (and willing) to use complex numbers.

• Back to the original root finding problem....

The solve command produced the roots but in the form of a sequence. If we wish to substitute these back into the polynomial to check the solutions, then the subs command must be used. i.e.

```
> subs(x=6, P);
```

The polynomial factorization we desire will be a product of the factors x-r, one for each root r, and the coefficient a[n] of the highest order term in the original polynomial, i.e.,

a[n] (x-r1) (x-r2)...(x-rn)

We construct this expression by Entering the following product command:

```
> rootList := [S];              # This converts  S  into a list.
> n := 5:  an := 3:
> an*product((x-rootList[k]), k=1..n);
```

Check to see if this really is the original polynomial by applying the expand command:

```
> expand(");
>
```

If this does not match the original polynomial then go back and check your work!

• RootOf.

The function RootOf represents all the roots of an equation.
The

allvalues command

is a convenient way to evaluate the RootOf command for polynomials. Enter the following command lines.

```
> P := 3*x^5-18*x^4-7*x^3+42*x^2-40*x+240;
> RootOf(P);
> allvalues(");
```

Explanation: RootOf is an inert command, meaning that it does nothing by itself, except identify what kind of variable it will be when the command is completed. The command allvalues, will complete the RootOf structure so it comes out with the desired result. The output should agree with what you obtained from your original execution of the solve command.

- Another factorization problem...with fsolve.

Here's another polynomial factorization problem (but with a new twist). Start by Entering the polynomial:

```
> P := x^7+6*x^6-3*x^5-x^4+2*x^3-x^2+5*x-1;
```

Try the factor command--what happens?

```
> factor(P);
```

Try the solve command -- what happens?

```
> solve(P=0);
```

For polynomials of higher degree, the solve command often returns RootOf(polynomial) where the variable x has been replaced by _Z . You may evaluate the RootOf with the allvalues command.

```
> allvalues(");
```

Note that the above results are numerical approximations. Maple is not always able to determine the exact roots of a polynomial of degree greater than four. When this is the case, Maple uses the fsolve command to find numerical solutions. The command fsolve applied directly to polynomials is a good option. Enter the following command.

```
> fsolve(P=0);
```

Notice, that Maple only returned three values. The fsolve routine will only return real roots unless the 'complex' options is included in the command. Then it will return the roots of the polynomial over the complex plane.

```
> fsolve(P=0,x,complex);
```

Now continue as we did with our first polynomial to achieve a complete factorization:

```
> rootList := ["]:
> k := 'k':  n := 7:   an := 1:
> an*product((x-rootList[k]), k=1..n);
> A := expand(");
```

If the expand command which checks your final answer produces some very small numbers (from roundoff errors), then apply the fnormal command to clean up the trash. The second argument in the fnormal command below gives the number of digits that you are rounding to. You should end up with the polynomial that you began with.

```
> d := 2:
> fnormal(A,d);
```

We comment here that fsolve is a very powerful tool and you are encouraged to find out more about it with the ?fsolve command.

The command fsolve can be applied to any polynomial to find the set of (approximate) roots, i.e., it can be successfully applied to more polynomials than solve. As an example, define p(x) to be some complicated 7-th degree polynomial and then apply both solve and fsolve to it. Enter your calculations below.

```
> p := ???;
>
> solve???;
>
> fsolve???;
```

How do the results of the two methods compare?

¤ 2. Operations with Rational Functions.

In this section we consider some of the algebraic operations which can be done on rational functions (quotients of two polynomials).

First we can take two (or more) rational functions and add them together, expressing the result over a common denominator. As an example, Enter the following combination of rational functions.

```
> 2/(1+3*x) + x/(1+x^2) + 1/(2-x);
```

Now combine them together with the normal command. Form of the command: normal(expr). Enter your command below.

• Maple Tip. Remember that " always stands for the output of the last executed Maple command. Hence normal(") is the quick way to apply normal to the last generated output.

```
> F := ???
```

Two other commands that are useful with rational functions are numer and denom. As you might guess, applying these commands to a rational function will yield the named portion of the fraction. Apply numer and denom to the rational function that resulted from the expand command above. Did you get what was expected? Form of the commands: numer(expr) and denom(expr). Enter your commands below.

```
>
```

Expand the denominator by using the expand command. Form of the command: expand(denom(expr)). Enter your command below.

```
> expand(denom(???));
> numer(F)/";
```

That was easy. However, what is really useful is how easy the reverse process is: taking a rational function and breaking it apart into pieces. This is the partial fraction decomposition which we were calculating by hand (in simple cases) earlier in the semester! Apply the command convert to the rational function you just obtained from the previous commands. Form of the command: convert(expr,parfrac,x). Learn more by using ?convert. Enter your command below.

```
>
```

Don't you wish that you could perform partial fraction decompositions that quickly? (Well, now you can...with a little help from Maple.)

Finally, another way to use normal is to cancel common factors in the numerator and denominator. As an example, apply the normal command to the following rational function: (x^2-1)/(x-1). Enter your commands below.

```
>
```

¤ 3. Limits and Derivatives.

• Limits.

Maple has a limit command which is easy to use. It also has the inert form with a capital L, which just prints out the limit in the usual form. Enter the following commands.

```
> Limit((x^2-4)/(x-2), x=2);
> limit((x^2-4)/(x-2), x=2);
```

We can also compute limits as the independent variable goes to infinity. Enter the following example.

```
> limit((x-2*x^2)/(x^2-4), x=infinity);
```

Now compute the limit of exp(-3*x) as x tends to infinity. Enter your commands below.

```
> limit(???);
```

• Derivatives.

Derivatives are straightforward in Maple, although there are a number of different notations in use. We will present derivative examples which work on the following function. Enter the following function definition command.

```
> f := x -> x^3;
```

First apply the diff operator to compute the derivative of the expression f(x). Form of the command: diff(expr, x), where expr is to be differentiated by x. Enter your commands below.

```
>
```

Next, apply the D operator to the function f.

```
>
```

What will we get if we enter diff(f(x),x,x)? Experiment with the commands below.

```
> diff(f(x),x,x);
> diff(f(x),x,x,x);
> diff(f(x),x$2);
> diff(f(x),x$3);
```

What do you think the expression diff(f(x),x,x,x) stands for? Place your answer below.

What do you think the expression diff(f(x),x$n) stands for? Place your answer below.

Now try the following commands.

```
> D(D(f));
> (D@D)(f);
> (D@@2)(f);
> (D@@3)(f);
```

• Maple Tip. For a constant function, say, h := x->6, Maple's output is h := 6.

What do you think the expression (D@@n)(f) stands for? Place your answer below.

Enter the following command as an introduction to the next section.

```
> D(f)(w);
> D(f)(2);
```

• Derivatives at a point.

How do we use these derivative operators to compute the derivative of f at a specific x value, e.g., at x:=3? In D this is easy: merely evaluate the derivative function at the desired point, i.e.,

```
> D(f)(3);
```

Using diff is more difficult. Because the result is in the form of an expression, the subs command can be used to substitute a value for x and thereby get a specific value.

```
> subs(x=3,diff(f(x),x));
```

• Maple Tip. In the above command, the expression diff(f(x),x) is first computed as 3*x^2, then the subs command substitutes x = 3 for x obtaining 27.

Compute the derivative of f(x) at x = 5 in two ways, one using D, and the other using diff. Enter your commands below.

```
>
```

The examples just given show a number of differences between D and diff for computing derivatives. Almost all of these differences are consequences of the fact that

D is applied to functions and diff is applied to expressions

Said more simply, D would be applied to f, while diff would be applied to f(x). To test your understanding of this difference, compute the derivatives of the following expressions in two ways: first with D, and then with diff:

arcsec(x^2)/x
x*exp(1-x^3)/(1+x^2)

Enter your commands below.

```
>
```

¤ 4. Antiderivatives and Indefinite Integrals.

• Antiderivatives.

Antiderivatives (or "indefinite integrals") are computed via the int command. Determine the indefinite integral for the function f(x)=x*sqrt(1+9*x^2) .
Form of the command: int(expr,x), where expr is to be integrated with respect to x. Enter your commands below.

```
> int ???
```

(What's missing from this answer that you know every general expression for an antiderivative must have? Leaving this off is simply the convention that Maple follows.) We now check this answer by differentiation, i.e., take the derivative of the expression just obtained--it should be the original function f(x). (Tip: Use the " symbol inside the derivative). Enter your command below.

```
>
```

The answer you just obtained might not look a lot like the function f(x) from which you started. If that is the case, then try to simplify your expression with the simplify command. (If simplify fails, then you can try some combination of the commands collect, combine, expand, factor, normal, convert, numer, and denom.

- Constants.

Any symbols which appear inside the integrate command (other than the variable of integration) are treated as though they are constants. As an example, determine the indefinite integral of the function f(x) = c/(a*x+b). Enter the following commands.

```
> int(c/(a*x+b), x);
```

- The Error Function. (Optional.)

Try to compute the following indefinite integral:

```
> integral(1);      # This is defined in your Lab 19 procedures.
```

$$2\,\frac{\int e^{\left(-x^2\right)}\,dx}{\sqrt{\pi}}$$

```
> f := sqrt(4/Pi)*exp(-x^2);
> int ???
```

You probably found this to be a surprising answer. Use Maple's ?xxxx command to get the full name of this new function. Enter your command below.

```
> ?erf
```

Want to know how this function is defined? Well, erf(x) is defined to be the value of the following definite integral:

```
> integral(2);
```

$$2\,\frac{\int_0^x e^{\left(-t^2\right)}\,dt}{\sqrt{\pi}}$$

Use Maple to compute this integral in order to check that it is indeed erf(x). (To obtain the command for the definite integral, Enter ?int.)

```
> ?int
```

erf(x) is defined by this definite integral (as in Lab 17). Hence we know the following facts:

- The derivative of erf(x) must be ???. (This derivative formula is a consequence of the ???.)
- The indefinite integral of

```
> f := t -> 2/sqrt(Pi)*exp(-t^2);
```

is given by ???.

- The definite integral

```
> integral(3);
```

$$2\,\frac{\int_a^b e^{(-t^2)}\,dt}{\sqrt{\pi}}$$

is given by ???. (This integral formula is a consequence of the ???.)

The erf function will appear in formulas for other integrals as well. For example, compute the following indefinite integral:

```
> int(exp(-(x^2)/2),x);
```

- Integration Practice.

Compute the indefinite integrals of the following three functions:

x*sin(5x^2), 1/(1+x^2), sqrt(1+x^2)

In each case check your answer by differentiation. (In one instance you will need to apply the Maple command simplify in order to make your derivative look like the original function.) Enter your commands below.

First Integral.

```
>
```

Second Integral.

```
>
```

Third Integral.

```
>
```

• More Integration Practice.

Let's consider a more substantial indefinite integration problem. Compute the indefinite integral of the function 1/(1+x^5). Enter your command below.

```
> F := ???
```

I'll bet you're glad you didn't have to do this integral by hand! How did Maple do the integration? Answer: It used an advanced version of partial fractions. To get a better feeling for the answer, determine its numerical version by using evalf. Enter your commands below.

```
>
```

Returning to the non-numerical, or symbolic indefinite integral, compute its derivative. Enter your computations below.

```
>
```

Gack! The output probably does not look like the function we started with! But appearances can be deceiving. Let's pull everything together over a common denominator with normal. Enter your computations below.

```
> f := ???
```

This should look better. To improve things even more we should expand both the numerator and denominator. Enter the cell below.

```
> expand(numer(f))/expand(denom(f));
```

Assuming the last step is correct, you are ready to factor the expression. Enter your command below.

```
>
```

You should be close to your goal. (Do you remember what that was?!) All you need to do is expand the denominator. Enter your command below.

```
> 1/expand( ??? );
```

Whew! Did you make it? If you have not arrived at the expression you started from, go back and check your computations.

¤ 5. Definite Integrals.

First let's compute the indefinite integral of y=log(x). Form of the command: int(expr,x), where expr is to be integrated with respect to x. Enter your computations below.

```
>
```

Now compute the definite integral of y=log(x) from x = 2 to x = 3. Form of the command: int(expr, x = a .. b), where expr is to be integrated with respect to x from x = a to x = b. Enter your computations below.

```
> ???
```

Notice that we obtain an exact answer (as opposed to a decimal, or floating point, number). Int always returns an exact answer (when it is able to return an answer at all). What theorem must int be using to compute values of definite integrals? Place your answer below.

To obtain a numerical answer for the definite integral, you can convert your last output to a floating point number by using evalf. Do this below. Enter your command below.

```
>
```

There is, however, a way to do the integration numerically from the beginning. By using the inert integration command Int and evalf, Maple will not symbolically evaluate the integral, it will start with a numerical approximation. Use this method to compute the definite integral of y=log(x) from x = 2 to x = 3 and compare the result to the answer you had obtained via int. Enter your command below.

```
>
```

A time trial is in order here between evalf(int(...)) and evalf(Int(...)). This will be easy to do by using the Maple command time(). This gives total CPU time in seconds since the start of the session. By taking the time() before and after a procedure and taking the difference, one can obtain the elapsed time of the procedure in seconds. Use this to time our two methods of definite integration evaluation for the function y=1/(1+x^5) over the interval x = 3 to x = 5. Enter the following commands.

```
> st := time():
>     evalf(Int(1/(1+x^5),x=3..5));
> time() - st;
> st := time():
>     evalf(int(1/(1+x^5),x=3..5));
> time() - st;
```

Which method is faster? Is the difference in speed large or small? If there is a significant difference in speed, how would you account for this behavior? Place your answers below.

¤ 6. Minimum Values from Plots. (Optional)

Enter the following function definition statement:

```
> s := x*sin(5*x^2);
```

We wish to plot the function over the interval from 0.3 to 1.3--Enter your command here:

```
>
```

We wish to determine rough coordinates for the minimum point that so clearly shows up in this graph. Maple plots allow us to do this very easily because you can use the mouse to find coordinates of any point in a graphics cell. To do this just click inside the graphic window and the coordinates will display at the bottom of the window.

On the basis of your coordinates, zoom in on the minimum point by constructing another plot statement, using a greatly reduced interval about the (approximate) x-coordinate of the minimum point. Enter your new plot here:

```
>
```

Continue the zooming in until you are confident of your x-coordinate to 4 decimal place accuracy. Compute the corresponding value of y. Enter your commands below.

```
>
```

¤ 7. Minimum Values from Calculus.

We start with the same function that was considered in the previous section.

In order to use calculus to determine the minimum value of the curve on the interval 0.3 <= x <= 1.3, we make use of the observation that at a minimum value, the derivative of the function must take on what value? ________. Hence we must first compute the derivative of s(x). Enter your command here:

```
>
```

Since we would like to find where the derivative equals ________, it seems natural to try and employ the solve function. Enter the appropriate solve command below:

```
>
```

Oh dear--that answer doesn't seem very helpful! The problem is that solve doesn't work well with many functions other than polynomials! Our only hope is to employ a numerical scheme to estimate the zeros of our derivative function. But wait...we know some very good methods for estimating zeros of functions. Use fsolve with an added range. Search for the root in the interval 0.3 <= x <= 1.3. Then compute the actual minimum value (i.e., the y coordinate value), and compare the answer with what you achieved by "zooming" in the previous section. Enter your computations below.

```
> fsolve(???,???);
> subs(???,s);
```

□ SUMMARY of COMMANDS

Here is a listing of all of the commands introduced in this Worksheet. (Do not, however, think this is an exhaustive list--there are plenty of other Maple commands that we could have discussed if more time was available). It might be wise to print a copy for convenient reference when using Maple at other times. Whenever you need information about these commands you have only to type ?CmdName, or look up the command in any Maple reference book.

- Algebraic Manipulations (mostly with polynomials).

```
factor(...);
solve(lhs=rhs,var);
fsolve(lhs=rhs,var);
RootOf(...);
expand(...);
simplify(...);
convert(...,type);
```

- Algebraic Manipulations with Rational Functions.

```
normal(...);
numer(...);
denom(...);
```

- Limits, Derivatives, Antiderivatives, and Integrals.

```
limit(expr,x=x0);
D(f)(x);
diff(f(x),x);
Int(f(x),x);
int(f(x),x);
Integrate(f(x),x = xmin .. xmax);
```

- Roots and Minimums for General Functions.

```
minimize(f(x), x, x=xmin .. xmax);
```

Additional Problems. (Optional)

1. Find all the real (i.e., not imaginary) roots of the following polynomial.

```
> p := -4 + 17*x - 10*x^2 + 21*x^3 - x^4 + 5*x^5 + 2*x^6;
```

2. Decompose the following rational function into a sum of rational pieces with denominators of degree no higher than 2. (This is *really* tough)

```
> r := (x^6-x^4+4*x^3-x+2)/(2*x^6+x^5-3*x^2+1);
>
>
```

3. Compute the derivative of the following expression.

```
> (1+x*exp(1-x^2)*cos(sqrt(x)) )/(2+sin(3*x)*exp(x^2));
```

4. Compute the indefinite integral of the following function, and then check your answer via differentiation.

```
> g := (x^3 +x)/(-30 + 35*x + 6*x^2 - 12*x^3 + x^5);
```

5. Compute the indefinite integral of the following function, and then check your answer via differentiation.

```
> g := (x^4 +2*x)/(-30 - 23*x + 6*x^2 - 3*x^3 + x^5);
```

☐

LAB 20: Trigonometric Substitutions

□ NAMES:

□ PURPOSE

In this lab you will use Maple to explore symbolic integration with an emphasis on the use of trigonometric substitutions involving the sine and tangent functions.

□ PREPARATION

Read Section 9.4, pages 547-552, in Calculus: Modeling and Application on substitution in integration.

□ LABORATORY REPORT INSTRUCTIONS

For this laboratory you are not required to submit a formal report. Rather you are asked to submit a copy of this Worksheet with the questions answered in the indicated locations. Only one submission will be made. Write in complete sentences and be sure the sentences connect together in coherent ways.

□ MAPLE PROCEDURES

```
> with(lab20);
> # If an Error is the output of the above line, see your lab instructor.
```

□ PRELIMINARIES.

¤ 1. The Method.

Let's summarize the basic ideas in the method of substitution in a way that we can apply it with Maple. Suppose we need an explicit formula for the indefinite integral of f(x), i.e., we need a function F(x) such that

```
> Int(f(x),x)=F(x)+C;
```

We find such an F(x) in a three step process. First we apply a substitution of the form x=g(u) to the integral Int f(x) dx. The result is that:

```
> Int(f(x),x) = Int(f(g(u))*D(g)(u),u);
```

However, f(g(u)) g'(u) equals F'(g(u)) g'(u) (why?),which by the Chain Rule equals (Fog)'(u). Thus our new integral Int f(g(u)) g'(u) du equals (Fog)(u)+C (why?). Our goal is therefore to find a substituting function x=g(u) such that the new integral can be evaluated, i.e., such that we can obtain an explicit formula for (Fog)(u). If we succeed, we perform the integration, the result of which is that:

```
> Int(f(g(u))*D(g)(u),u) = (F@g)(u);
```

However, we wish a formula for F(x), not (Fog)(u), and thus we back-substitute u=g(-1)(x), (g inverse of x) into our formula for (Fog)(u). The result is that:

```
> (F@g)(g(-1)(x)) = F(x);
```

The three steps in this method are visually summarized in the following diagram. This is how substitution will be done in the Project.

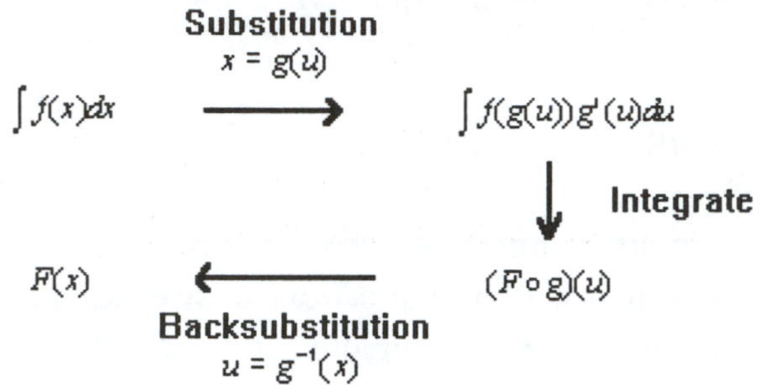

Substitution
$x = g(u)$

$\int f(x)dx \longrightarrow \int f(g(u))g'(u)du$

$\downarrow$ **Integrate**

$F(x) \longleftarrow (F \circ g)(u)$

Backsubstitution
$u = g^{-1}(x)$

¤ 2. Substitution with Maple.

Suppose we are trying to determine the integral for a function f(x). We wish to transform the integral Int f(x) dx into the (hopefully simpler) Integral, Int f(g(u))g'(u) du. Maple allows us to take any function g(u) and make the desired transformation easily by using a command called changevar. This command will substitute g(u) into an integral wherever x occurs. Study the following example to see what we mean.

• Step One: Substitution.

```
> changevar(x=g(u),Int(f(x),x),u);
```

As you can see, our command is to "substitute x=g(u) into the Integral, Int f(x) dx," and the result is the Integral, Int f(g(u))g'(u)du, as desired. In this manner we can test a function g(u) and easily determine whether or not the resultant substitution is useful or not (i.e., produces an expression that we can more easily integrate).

Once we have settled on a particular function g we then integrate the remaining terms via Maple's Integrate command. Enter the following commands:

• Step Two: Integration.

```
> subs(f(g(u))*diff(g(u),u)=diff((F@g)(u),u),");
>   # This step is only needed in general case.
> value(");
```

Once the integration is done, we now have only to solve the equation x=g(u) for the variable u and substitute it into our indefinite integral. Enter the following commands:

• Step Three: Back-substitution.

```
> subs(g(u)=x,"); # This step only needed in general case
> #solve(g(u)=x,u); # These steps NEEDED in specific case
> #subs(u=","");
> simplify(");
```

We thus are left with F(x), the function (indefinite integral) that we wanted to compute.

Study the commands we have used in this section very carefully--they constitute the basic set of procedures that we will apply to all of the examples in this lab. For your convenience we collect these commands below. (Do not Enter them at this time!)

```
f := 'f':  g := 'g':

f := x -> ???;

g := u -> ???;

Int(f(x),x);

changevar(x=g(u),",u);        # Step one:   Substitution

value(");                     # Step two:   Integration

solve(g(u)=x,u);

subs(u=",""");                # Step three:   Back-substitution

simplify(");
```

¤ 3. Trigonometry with Maple.

We will use two Maple commands that apply basic trigonometric simplifications (e.g., sin(-x) -> -sin(x)) to an expression: expand(expr) and combine(expr,trig). Here are examples. Enter the following commands.

```
> cos(x)^2;
> combine(",trig);
> expand(sin(2*x));
```

□ PROJECT

¤ 0. Some Starting Integrals.

We need to determine some integral formulas to be used in the subsequent examples on integration by substitution. We first determine the integral of

f(x)=cos(x)^2.

Enter the appropriate function definition statement.

```
> f := x -> ???;
```

Enter the appropriate Maple command to integrate f(x).

```
> ???
```

How did Maple compute this integral? Well, it used an important trigonometric identity which expresses the square of the cosine in terms of a non-squared cosine. Apply one of the new trigonometry commands (Section 3 of the Preliminaries) to f(x) to obtain the desired formula. Enter your command below.

```
> ???(???, trig);
```

You should be able to integrate each term in thi new representation of f(x) by hand. However, since the computer is in front of you, use Maple to compute the integral of the expression. Enter your command below.

```
>
```

Your final answer should agree with the one that you produced originally just using the Integrate command. If not, then check your work.

****

We next wish to consider the integral of

f(x) = sec(x).

First check what Maple has to say about the integral. Enter your function definition and integrate commands below.

```
> f := x -> ???;
>
```

Check that this is indeed an antiderivative for sec(x) by differentiation. (You may need to apply a simplify command to the result of the differentiation.) Enter the appropriate command.

```
>
```

Note. There is a direct method for obtaining the integral formula

Int sec(x) dx = log (tan(x)+sec(x)) + C

The method is just a flat-out clever trick: multiply the integrand sec(x) by (sec(x)+tan(x))/(sec(x)+tan(x)). The numerator will then turn out to be the derivative of the denominator, which allows for the evaluation of the integral. We'll leave the details to you.

****

¤ 1. The First Integral.

We now use substitution to compute some non-trivial integrals. We first consider integrating the function

f(x) = sqrt(25-x^2);

First use Maple to compute the integral, and then compute the derivative of your answer. Enter your computations below.

```
> f := x -> ???;
> int(f(x),x);
```

Now compute the derivative of your answer, to see if you get back to the original function. Enter your computations below.

```
>
```

Hmmm.... As with the previous integral, we need some simplification techniques. Try simplify. Enter your computations below.

```
>
```

****

Now we attempt a more careful analysis of what Maple is doing to compute this integral. We claim that Maple saw the term sqrt(25-x^2) and realized that a good substitution would be x = 5*sin(u). The reason? Because 25-25*sin(u)^2 equals 25*cos(u)^2. Let's do the computations in the manner we outlined in the Preliminaries section.

We begin with Step One: substitute x = 5*sin(u) into f(x) dt(x). Place your command below.

```
> changevar(???,???,???);
```

Before proceeding on to Step Two (Integration) we need to simplify the output from the substitution. Start by applying the replacement rule

sin(u)^2 = 1-cos(u)^2

followed with another replacement rule,

sqrt(cos(u)^2) = cos(u).

Enter your commands below.

You should have obtained a simple (and familiar) result--you worked with this integral in the Preliminaries section. We now can proceed with Step Two of the Method of Substitution: Integration. Enter your commands below.

```
> value ???
```

We're now at Step Three: Back-substitution. We solve x = 5*sin(u) for u and substitute the result into the formula just obtained for the integral. Enter your commands below.

```
>
```

You should have a perfect match with the original integral!

****

Let's try a definite integral with the same integrand. Compute the definite integral

int(sqrt(25-x^2),x=2..4);

in three different ways:

(1) by using the indefinite integral that you just computed,
(2) by using Maple's int command, and
(3) by using Maple's evalf(Int) command. Compare your answers.

Place your commands below.

```
>
```

****

¤ 2. The Second Integral.

Now we turn our attention to computing the integral of the function

f(x) = 1/sqrt(4-x^2)

As done for the function in Section 1, compute the integral using int and check your result by differentiation. (You may need to use an algebraic simplification rule at some point.) Place your commands here:

```
> f := x -> ???;
> int(???);
```

****

As done for the function in Section 1, we wish to recompute the integral using substitution. Begin this process by completing Step One: choosing an appropriate substitution function and carrying out the substitution. Enter your commands below.

```
>
```

Before proceeding on to Step Two (Integration) we need to simplify the output from the substitution. Apply the replacement rules

sin(u)^2 = 1-cos(u)^2 , and
1/sqrt(cos(u)^2) = 1/cos(u)

followed by a simplify command.

```
> subs(???,");
> subs(???,");
> simplify(");
```

You should have a very simple integral in front of you. We now can proceed with both Step Two (Integration) and Step Three (Back-substitution). Enter your commands below.

```
>
```

Does your final answer match the formula above? Place your answer below.

****

Compute the definite integral

Int(1/sqrt(4-x^2),x=0.24..1.73)

in three different ways, and compare your answers. Place your commands below.

```
>
```

****

¤ 3. The Third Integral.

Now we turn our attention to computing the integral of the function

f(x) = 1/sqrt(4*x^2 + 1)

As done for the function in Section 1, compute the integral using Integrate. Place your commands below.

```
> f := x -> ???;
> int(f(x),x);
```

Hmmm.... You answer will include an arcsinh (the inverse of the "hyperbolic sine function"), a function that we have not studied. However, there is another form of the answer which does not use hyperbolic functions and will be more useful in our focus on substitution. To get this other form of the answer we use the Maple command convert.

```
> convert(",expln);
```

You should now have an answer for our integration problem that involves a log and a square root function. Check that this answer is indeed correct by differentiating it and comparing it to the integrand of the original integral. Do you obtain what is desired? (You might need to simplify the output of the differentiation before you can answer the question.) Enter your commands below.

>

****

Again, as done for the function in Section 1, we wish to recompute the integral "manually," using an appropriate substitution. In the current situation you should look for a substitution involving the tan and sec functions. The reason: there is an important identity between tan(x)^2 and sec(x)^2. What is this identity?

sec(x)^2 - 1 = tan(x)^2

We begin with Step One: Substitution. Use the identity just obtained to pick the appropriate substitution x = ??? for the integral above. Then simplify your answer with the command simplify, followed by the replacement rule sqrt(1+tan(u)) = sec(u). The result should be a fairly simple differential whose integral we know from previous work! Place your commands below.

>

Ah ha! We have reduced ourselves down to needing the integral of a secant function, but we worked out the integral of sec(x) in Section 0. Thus the integral of what is before us now can be written down in terms of int(sec(u),u). In this way Step Two of the Method of Substitution (Integration) is completed without further computation. Enter the desired integral below (an appropriate multiple of Intsec(u)).

>

You are now ready for Step Three: Back-subtitution. We solve x=??? for u and substitute the result into the formula just obtained for the integral. Enter your command below.

>

The answer you have obtained should match the answer we obtained earlier from int (plus simplifications).

¤ 4. The Fourth Integral.

Consider the following function:

f(x) = 1/sqrt(9*x^2 + 6*x + 2)

As done for the previous integrals, use Maple to compute the integral of f(x), and then check your answer via differentiation. (Note: You'll obtain another arcsinh function as in Section 3--get rid of it in the same way as you did in Section 3.) Enter your commands below.

Again, as done for the previous integrals you will pick an appropriate substitution and evaluate the integral via your substitution. However, in this case the appropriate substitution may not be so self-evident: unlike the previous examples, the polynomial expression in f(x) has a first order term, i.e., 6*x. This problem can be eliminated by completing the square. We now provide you with a routine which takes any quadratic polynomial and completes the square. Enter the following function definition command and the subsequent example.

```
> CTS := proc(p,x)
> local a,i,P;
>   P := unapply(p,x);
>   for i from 0 to 2 do
>     a[3-i] := (D@@i)(P)(0)/i!;
>   od;
>   (sqrt(a[1])*x+a[2]/(2*sqrt(a[1])))^2-(a[2]^2-4*a[1]*a[3])/(4*a[1]);
> end;
> CTS(x^2+2*x+2,x);
> simplify(");
```

Apply the CTS to the polynomial portion of f(x). Use the result to decide on an appropriate substitution to use to evaluate the integral and list your choice here. Enter your commands below.

```
> CTS ???
```

Now carry out Steps One, Two, and Three of the Method of Substitution. The steps here are very much like those done for the third integral. Enter your commands below.

```
> solve ???
> g := u -> ??? ;
> ???
> subs(???,");
> changevar ???
> subs(???,");
> subs(???,");
> value(");
> ???
> subs(u=",""");
> simplify(");
```

Does your final answer agree with what was obtained directly from Integrate (after simplification)? If not, then check your work. Place your answer below.

□

LAB 21: Normally Distributed Data

□ NAMES:

□ PURPOSE
In this lab we consider a set of normally distributed data, developing techniques to analyze the data and to handle simple statistical questions. The mean and standard deviation of the data set are investigated, and the statistical significance of the bell curve is revealed.

□ PREPARATION
Read Section 10.4, pages 639 - 649, in Calculus: Modeling and Application.

□ LABORATORY REPORT INSTRUCTIONS
Note: Follow your instructor's instructions if they differ from those given here.

For this laboratory you are not required to submit a formal report. Rather you are asked to submit a copy of this Worksheet with the questions answered in the indicated locations. Only one submission will be made. Write in complete sentences and be sure the sentences connect together in coherent ways.

□ MAPLE PROCEDURES

```
> with(lab21);
> # If an Error is the output of the above line, see your lab instructor.
```

□ PRELIMINARIES

¤ 1. Symmetric Difference Quotients and Numerical Derivatives.

Suppose s is a quantity which depends on a variable t, and we have measurements for s at the t-values t1<t2<t3. How could we numerically estimate the derivative of s at t2? We will need to make such estimates later in this lab.

Our first thought, since we are working with data at a number of discrete t values, would be to approximate the derivative at t = t2 by the difference quotient

Deltas/Deltat = (s(t3)-s(t2))/(t3 - t2)

We claim, however, that a better estimate to the derivative is generally obtained by using what we shall call the symmetric difference quotient

(Deltas/Deltat)sym = (s(t3)-s(t1))/(t3 - t1).

As you can see, to approximate the derivative at t2 via symmetric difference quotient (SDQ) we use the function points to either side of t2, i.e., (t1, s(t1)) and (t3, s(t3)).

To convince you that, in most cases, the symmetric difference quotient is indeed the better approximation to the derivative, we present some graphical evidence. Enter the following commands.

```
> func := t -> t^3;  t2 := 1.0;  run := .5;
> sdqPlotter(func, t2, run);
```

Explanation: The function func(t) = t^3 is plotted for the interval 0.5 <= t <= 1.5 . Three points are shown as dots, corresponding to t1=0.5, t2=1.0, and t3=1.5. Delete the incorrect choices for each question concerning the center point t2.

- The difference quotient is the slope of the (red, blue, or dashed) line.
- The symmetric difference quotient is the slope of the (red, blue, or dashed) line.
- The derivative is the slope of the (red, blue, or dashed magenta) line.
- The derivative is best approximated by which quotient? ("regular" or symmetric)

Now this is just one example--perhaps the outcome is different for most other functions! Well, one of the powerful features of Maple is that you can generate as many examples as you like! The command sdqPlotter is designed to generate LOTS of examples--all you need to do is make the desired changes. You can change the function func(t), the t-value t2, and the distance "run" between t2 and t1 (and between t2 and t3). So that you don't destroy the good example generated above, a copy of the input statements are given below for your alteration. Try some of your own examples, leaving the "best" one you are able to generate as an additional illustration of the relationships between the derivative, the difference quotient, and the symmetric difference quotient. Enter a few examples of your own below.

Suggestion: It is instructive (and amusing) to try and find aberrant situations in which the symmetric difference quotient is worse than the (ordinary) difference quotient as an approximation to the derivative. Such examples are interesting, however, only if run is small.

```
> func := ???: t2 = ???: run = ???:
> sdqPlotter(func, t2, run);
```

For the convenient use of symmetric difference quotients in the numerical approximation of derivatives, we have written a special command, nD ("numerical derivative"). This command will approximate the derivative of any function by a symmetric difference quotient; the default value of the run is .0001, although any run can be specified by use of the option sdqDelta. (More information on nD can be gotten with the usual "?" command.) Enter the following examples of nD.

```
> f := x -> x^3:
> nD(f);
> nD(f)(x);
> simplify(nD(f)(x));
```

Is nD(f) a good approximation to the derivative of x^3? Why? Place your answer below.
Answer:

Here are examples using a version of ND constructed for evaluating a specific x=x0 value. Enter the following examples of ND.

```
> nD(x ->x^5)( 2);
> nD(sin)(0);
```

Are these values good approximations for the specified derivatives? Place your answers below.
Answer:

¤ 2. The Data Set.

The following is a data set consisting of the heights (in inches) for 100 women. The lab will be concerned with analyzing this data set. Enter the data list Height.

```
> Height :=   [ 61.28, 57.20, 63.85, 62.61, 63.88, 62.67, 59.36, 61.48,67.61, 67.03, 63.93, 61.5
> 4, 61.86, 59.62, 63.18, 63.49, 60.93, 64.74, 64.73, 64.55, 68.82, 57.28, 61.42, 61.66, 59.48, 61.
> 27, 64.01, 57.17, 64.50, 62.24, 62.48, 59.96, 61.39, 60.48, 59.69, 63.57, 58.18, 56.00, 62.53, 62
> .38, 61.38, 58.88, 58.76, 61.94, 62.49, 60.24, 62.01, 57.50, 62.92, 64.89, 63.68, 62.75, 67.50, 6
> 4.03, 64.71, 60.73, 59.80, 65.61, 65.86, 59.98, 57.97, 64.03, 61.76, 63.83, 62.92, 64.28, 62.62,
> 62.68, 66.74, 62.65, 67.16, 64.94, 62.04, 62.07, 66.07, 63.91, 59.73, 61.27, 58.92, 62.12, 58.74,
>  63.07, 63.78, 63.45, 66.58, 64.40, 59.98, 61.34, 65.25, 65.25, 63.39, 61.35, 62.15, 62.35, 61.78
> , 60.39, 60.57, 62.57, 64.34, 63.69]:
```

□ PROJECT

¤ 1. Problems to be Addressed.

In this section you will need the data set Height from the Preliminaries.

We will analyze a data set Height consisting of the heights (in inches) for 100 women. This data set was defined and Entered in the Preliminaries section of the lab. Display the data set for closer examination. (To do this, simply type and Enter the word Height.) Display the data set below.

```
> Height;
> nops(Height);
```

What can you say about the data just by visual examination of the numbers, i.e., without doing additional calculations? Place your observations below.

Answer:

While some useful observations can be made in this way, our ability to make sense out of data simply by "examination" is usually limited. Perhaps seeing a scatterplot would be helpful. Plot the data set Height by Entering the following command. (The command ListPlot was written for this lab.)

```
> ListPlot(Height);
```

What can you say about the data just by examining the scatter plot, i.e., without doing additional calculations? Place your observations below.
Answer:

You will probably agree that nothing particularly sophisticated has yet been determined about the data. For example, based on what has been done thus far, could you answer a question such as "What is the probability that a woman's height falls between 60 and 61 inches?" Probably not (at least not without further computations). These deficiencies yield the following two problems for our consideration:

- Problem 1. Can we make sense, or derive some order, out of the data? In particular, is there a useful model for the data that reveals this order?

- Problem 2. Can we compute the probability that a randomly chosen height will fall into a specified range of height values? And can we compute this efficiently?

The remainder of the lab will focus on addressing these two problems.

¤ 2. The Distribution Function F(t).

- In this section you will need the data set Height from the Preliminaries.

An object which is essential in helping us with Problems 1 and 2 is the Distribution Function F(t) for the data set under consideration. F(t) is merely the fraction of the total number of data points with values less than t, i.e., the distribution function F(t) can be viewed by activating the following line.

```
> `F(t)` = (`Total number of data points with value < t`)/ (`Total number of data points`);
```

$$F(t) = \frac{\text{Total number of data points with value} < t}{\text{Total number of data points}}$$

Phrased in the language of statistics, F(t) is the probability that a randomly chosen data point will have a value which is _____??_____.

One way to construct the distribution function F(t) in Maple is as follows.

1. Sort the height data into increasing order.
2. For each t, count the heights m(t) which are less than t (easy since the heights are sorted).
3. The distribution F(t) is now just m(t)/m, where m is the total number of data points.

Let's carry out this procedure. Sorting a data list is done using the command Sort. Enter the following command.

```
> sortedHeight := sort(Height);
```

Now count the number of heights which are less than t=60, and then determine the value of the distribution function F(t) at t=60. What is the probabilistic meaning of this number? Place your answer below.
Answer:

Since we will need to compute distribution functions for other data lists, we provide a command DistributionFunction which carries out the procedure just described on any sorted data list. It is used as shown in the next cell, which defines F(t) to be the distribution function for the Height data. Enter the following commands which define the distribution function F(t).

```
> F:= t -> DistributionFunction(sortedHeight,t);
```

Check it by activating the following.

```
> F(60);
```

Plot the distribution function F with the following command. It will be a little slow and the single quotes ' 'around F(t) are necessary.

```
> FPlot:=plot('F(t)',t=56..68):";
```

Using the distribution--Indicate the probabilistic meaning of each of the following four expressions and, if appropriate, determine the numerical value. Place your answers below each expression:

```
> F(61.3);
```

Meaning:

```
> F(65.8);
```

Meaning:

```
> F(65.8)-F(61.3);
```

Meaning:

F(b)-F(a); # for a < b
Meaning:

The Problems Reviewed--Let us recall the problems listed at the end of the previous section and determine how much progress we have made with them.

• Problem 1. Can we make sense, or derive some order, out of the data? In particular, is there a useful model for the data that reveals this order?

What progress has been made on this problem? What has yet to be done?

Progress on Problem 1:

• Problem 2. Can we compute the probability that a randomly chosen height will fall into a specified range of height values? And can we compute this efficiently?

What progress has been made on this problem? What has yet to be done?

Progress on Problem 2:

¤ 3. The Mean and Standard Deviation.

• In this section you will need the data set Height from the Preliminaries.

An amazing amount of statistical information can be deduced for many data lists by knowing just two numbers: the mean and the standard deviation. The importance of these numbers will be revealed in later sections. In this section we merely define and compute the numbers.

• The mean (μ, "mu") of a data set is the average of all the values. For a data list [h1, h2,..., hn], the precise definition is seen by activating the following command line.
mean =

```
> mu := (1/n)*Sum( h[k], k=1..n);
```

The mean should visually correspond to the "center" of the data list. From the scatter plot given in Section 1, make a rough estimate for the mean μ of the data list Height. Place your answer below.
Estimate:

It is not difficult to use basic Maple commands to compute the mean for a data list. However, for convenience we have a command Mean that can be applied to any data list. Enter the examples below.

```
> Mean([2, 3, 4, 5]);
```

```
> Mean([4, 9, 16, 25]);
```

```
> mean := Mean(Height);
```

This last example sets "mean" to be the Mean of the data list Height. How good was your original estimate for this quantity?

• The standard deviation "sigma" of a data list measures the amount of deviation from the mean μ--the larger the sigma, the greater is the distance from the mean for a typical data value. The Greek letter sigma can be seen by activating the following line.

```
> 'sigma';
```

For the rest of the lab we will use the letter s in place of sigma since we do not have access to the Greek letter sigma in this text material. For a data list [h1, h2,..., hn], the precise definition is:

standard deviation =

```
> s :=sqrt((1/n)*Sum((h[k]-'mu')^2, k=1..n));
```

$$s := \sqrt{\frac{\sum_{k=1}^{n} (h_k - \mu)^2}{n}}$$

The standard deviation s measures the "spread" of the data list. To be more precise, approximately 70% of the values of a "normal" data list should fall within one standard deviation of the mean, i.e., within the range μ-s <= h <= μ+s (a justification for this non-obvious fact will be given later). From the scatter plot given in Section 1, make a rough estimate for the standard deviation s of the data list Height. Place your answer below.
Estimate:

For convenience we have a command StandardDeviation that can be applied to any data list. Enter the examples below.

```
> StandardDeviation([2, 3, 4, 5]);
> StandardDeviation([4, 9, 16, 25]);

> sd := StandardDeviation(Height);
```

This last example sets sd to be the standard deviation of the data list Height. How good was your original estimate for this quantity?

◻ 4. Normalized Data.

• In this section you will need the data list Height from the Preliminaries.

One way to arrive at a useful model for a data list is to use the mean and the standard deviation to transform the data list into its normalized form. The normalized data list is obtained by taking each data value, subtracting the mean μ, and then dividing by the standard deviation s. Thus if the data list

data := [d1, d2, ..., dn], then the

normalized data list := [(d1 - μ)/s, (d2 - μ)/s, ... , (dn - μ)/s].

To understand the important properties of the normalized data list, define a data list "data" below, then Enter the two subsequent regions. The data will be plotted, its mean and standard deviation will be computed, and then the same will be done for the normalized data set. Complete and Enter the following example.

```
> data := ???:
> ListPlot(data);
> meandata := Mean(data): lprint(`mean =`, meandata);
> sddata := StandardDeviation(data): lprint(`standard deviation =`, sddata);
```

Now we do the same for the normalized data.

```
> normdata := map(x->(x- meandata)/sddata,data):
>     # This normalizes the data.
> ListPlot(normdata);
> meannormaldata := fnormal(Mean(normdata),4):
>     # fnormal rounds off decimals.
> lprint(`mean of normalized data =`, meannormaldata);
> sdnormaldata := fnormal(StandardDeviation(normdata),4):
> lprint(`standard deviation of normalized data =`, sdnormaldata);
```

On the basis of your example, make a conjecture about an important property of a normalized data set. Test your conjecture by changing the example data set and re-executing the commands above. Does your conjecture still hold true? Place your answers below.
Answer:

Using the definition for the normalized data set, give a justification for your conjecture. Place your answer below.
Answer:

For use in the next sections, define normHeight to be the normalized Height data list. Enter the following commands. We have written a command Normalize for this lab which normalizes any list of numbers. Check ?Normalize for more details.

```
> mean := Mean(Height);
> sd := StandardDeviation(Height);
> normHeight := Normalize(Height):
> Mean(normHeight);
> StandardDeviation(normHeight);
```

• Maple Tip. Computers usually have round-off errors, so if the results you get aren't exactly what you expect, it may be because of round-off errors. You may be happier to use the fnormal command as we did above. For more information try the ?fnormal help command.

```
> fnormal(Mean(normHeight),5);
> fnormal(StandardDeviation(normHeight),5);
```

¤ 5. The Normalized Distribution Function normF(t).

• In this section you will need the data set Height from the Preliminaries and the corresponding normalized data set normHeight, defined in the previous section.

Given our construction of the normalized height data set normHeight, we now return to consider the distribution function F(t). For convenience, enter the commands below to recompute and re-plot the function. Enter the following commands.

```
> sortedHeight := sort(Height):
> F := t -> DistributionFunction(sortedHeight,t);
> FPlot;
```

We now compute the normalized distribution function normF(t), i.e., the distribution function for the normalized data normHeight, in the same way that the original distribution function was defined. Enter the following commands.

```
> sortedNormHeight := sort(normHeight);
> normF := t -> DistributionFunction(sortedNormHeight,t);
> normFPlot := plot('normF(t)',t=-2.6..2.6):";
```

Comparison between the two distribution functions--How does the plot of the normalized distribution function normF(t) compare with the plot of the (unnormalized) distribution function F(t)? Carefully describe the similarities and the differences. Place your answer below.
Answer:

¤ 6. The Density Function f(t).

• In this section you will need the data set Height from the Preliminaries and the distribution function F(t) as defined in Section 2.

For convenience, enter the commands below to recompute the distribution function F(t). Enter the following commands.

```
> sortedHeight := sort(Height):
> F := t -> DistributionFunction(sortedHeight,t);
```

* * * * *

We need to introduce one last new concept: the density function of a data list. We motivate its definition as follows.

Let F(t) be the distribution function for a data list. From Section 2 we know that the quantity F(b)-F(a) gives the probability that a value chosen randomly from the data list will fall in the interval a <= t <= b. Hence we would like to have an efficient method for computing F(b)-F(a). Recall, however, that if f(t) is the derivative of F(t), then the First Fundamental Theorem of Calculus gives the following formula for F(b)-F(a): Activate the following line.

```
> t:='t': `F(b)-F(a)` := Int(`f(t)`, t=a..b);
```

We call f(t) the density function of our data list. Thus, if we know the density function f(t), we can compute all probabilities F(b)-F(a) merely by definite integration.

Computing the density--We will numerically approximate the derivative (density function) f(t) of F(t). This will be done with the symmetric difference differentiation command sdqD, which approximates derivatives with symmetric difference quotients. To make these computations easy, we have set up a "seq" command which takes a sampling of the domain of the distribution function F. This sampling will be called "Partition" and will be of the form

Partition := [a, a+deltat, a+2*deltat, a+3*deltat, ...].

Then we take the distribution function F of each entry in Partition,

FPartition := [F(a), F(a+deltat), F(a+2*deltat), ...].

Applying the Zip command to Partition and FPartition yields a function sequence whose terms are of the form [t, F(t)] to which we can apply the sdqD differentiation command which produces a list of density function (derivative) values [t, f(t)]. Thus a data point t in the set "Partition" goes through the following path.

t -> F(t) -> [t,F(t)] -> [t,f(t)].

These points are then plotted to give the graph of the density function. Enter the following commands.

```
> deltat := .25:  # The step size between t values.  Try different values.
> Partition := [seq(56+i*deltat, i=0..13/deltat)];
```

The Fof command has been written so that Fof(List) returns a list of the F values of the members of List. Use ?Fof.
Activate the following commands.

```
> FPartition := Fof(Partition);
> DistributionValues := Zip(Partition, FPartition);
> DensityValues := sdqD(DistributionValues);
> densityGraph := plot(DensityValues, color=COLOR(RGB,.1,.7,.1),thickness=3):";
```

Does your (approximate) density function graph look unpleasantly jagged? If so, this is a result of the "jagged" nature of the original distribution function F(t). The "jaggies" can be minimized by experimenting with the size of the increment "deltat" used to sample the data which in turn is used to compute the symmetric difference derivatives, sdqD, of the data set Fdata. Experiment with values of "deltat" in the previous set of commands.

The Bell Curve--As you should see, the graph of the density function f(t) has a familiar "bell-shape." We have seen exponential functions that have this shape--maybe we could approximate (i.e., "model") the jagged density function f(t) by an expression whose graph is a smooth bell curve! The answer to this question is Yes, as we shall soon see.

¤ 7. The Normalized Density Function normf.

• In this section you will need the graph densityGraph from the previous section and the normalized distribution function normF as defined in Section 5.

```
> sortedNormHeight := sort(normHeight):
> normF := t -> DistributionFunction(sortedNormHeight,t);
```

As done in the previous section for the (unnormalized) data set, we will now compute and plot approximate values of the density function normf for the normalized data set normHeight as defined in Section 4. As before, we have a seq command which takes the normalized distribution function normF(t) and produces a list of density function values [t, normf(t)] by using sdqD. These points are then plotted to give the graph of the density function. Enter the following commands.

```
> deltat := .2:  # The step size between t values.  Try different values.
> normPartition := [seq(-2.5+i*deltat, i=0..5/deltat)]:  # Partitioning the interval [-2.5, 2.5] .
```

The normFof command has been written so that normFof(List) returns a list of the normF values of the members of List. Use ?normFof.
Activate the following commands.

```
> FnormPartition := normFof(normPartition):
> normDistributionValues := Zip(normPartition, FnormPartition):
> normDensityValues := sdqD(normDistributionValues):  # defines normf
> normDensityGraph := plot(normDensityValues, color=COLOR(RGB,.5,.4,0),
>       thickness=3):";  # graph of  normf
```

If necessary, adjust the value of "deltat" in the normDensityGraph commands until a relatively decent density curve is obtained.

Comparison with the unnormalized case--In what ways is the graph of the normalized density function similar to the graph of the (unnormalized) density function? In what ways do the graphs differ? (The following command places both graphs into the same plot.) Enter the following command, then place your answers below.

```
> plots[display]({normDensityGraph, densityGraph});
```

Answer:

Amazing Fact: Many data sets from the real world will produce essentially the same normalized density graph as obtained for Height! That is the primary reason why normalizing a data set is such a useful procedure.

¤ 8. A Model for the Density Function.

• In this section you will need the graphs densityGraph and normDensityGraph, as well as mean and sd, the mean and standard deviation of the data set Height.

For convenience, enter the commands below to recompute mean and sd. Enter the following commands.

```
> mean := Mean(Height); sd := StandardDeviation(Height);
```

* * * * *

It's time to return to the question we raised at the end of Section 6--perhaps we can approximate (i.e., "model") the jagged density function f(t) by an expression whose graph is a smooth bell curve!

The answer is Yes. We first consider the normalized density function.

As we will see, a multiple of the function exp(-t^2/2) will approximate the normalized density function quite well. We will determine the multiplicative constant experimentally, i.e., by finding the best fit with our normalized density graph.

We computed the graph of the normalized density function in the previous section--it was called normDensityGraph. This next command puts the graph of the function c*exp(-t^2/2) on top of normDensityGraph for any user specified value of c. Determine the best c value! Enter the following commands for as many c values as necessary.

```
> c:= 1.0;
> plots[display]({normDensityGraph, plot(c*exp(-t^2/2), t=-3..3)});
```

With the value of c that you just found, you have constructed the following model for the normalized density function. Enter the following command.

```
> normf := t -> c*exp(-(t^2)/2.);
```

Amazing Fact: Many data sets from the real world will produce essentially the same normalized density graph as obtained for Height, and hence the model for the normalized density graph will still be the function normf(t). That is why this function is so important in the study of statistics--it arises over and over again!!

But what about the original (unnormalized) density function? Well, if t is an (unnormalized) height value, then we claim that the (unnormalized) density f(t) should equal the normalized density divided by sd at the point (t-mean)/sd, i.e.,

f(t) = (1/s) normf((t-mean)/s)

Hmmm.... This is hardly a transparent claim!! How can we justify it? Well, if you examine our work, you will see that

F(t) = normF((t-mean)/s)

Differentiating both sides of this equation will prove our claimed relationship between f and normf. Hmmm.... we bet you still don't believe us.... Then we'll show that the claim is correct!

We define f(t) to be the claimed model for the (original) density function. Enter the following definition for f(t):

```
> f := t -> (1/sd)*normf((t-mean)/sd);
```

We now place the graph of f(t) on top of densityGraph, the graph of the original density function computed in Section 6. You decide how good the approximation is--seeing is believing! Enter the following plots[display] command:

```
> plots[display]({densityGraph, plot(f, 55..70)});
```

Does it appear that our function f(t) is a good model for the density of our data set Height? Place your answer below.
Answer:

¤ 9. ...and back to our Original Problems.

• In this section you will need the density function f(t) and the distribution function F(t) for the data set Height.

Remember how the distribution function F(t) and the density function f(t) are related: the density f(t) is the derivative of the distribution F(t), and the 1st FTC (Fundamental Theorem of Calculus) gives us that

```
> `F(b)-F(a)` := Int(`f(t)`, t=a..b);
```

$$F(b)-F(a) := \int_a^b f(t)\, dt$$

for all values of a and b. Remember why the expression F(b) - F(a) is important: it is the probability that a value t chosen at random from Height will be in the interval a <= t < b. We now have a new way to approximate this quantity: numerically evaluate the corresponding definite integral of our density model function f(t). Let's see how accurate our approximations will be, and further, let's see how fast the computations are. Enter the following commands; then Enter some of your own.

```
> F(61)-F(57);
> int(f(t), t=57..61);
>
```

On the basis of your computations, what can you say about the speed of our two methods, and the accuracy of the two methods? How do you think your answers would change if we had 1000 height measurements instead of just 100? Place your answers below.
Answer:

Now we finally return to our problems from Section 1:

• Problem 1. Can we make sense, or derive some order, out of the data? In particular, is there a useful model for the data that reveals this order?

Summarize what we have done to handle this problem:
Answer:

• Problem 2. Can we compute the probability that a randomly chosen height will fall into a specified range of height values? And can we compute this efficiently?

Summarize what we have done to handle this problem:
Answer:

¤ 10. Additional Questions to Ponder. (Optional.)

We first note that the appearance of the function c*exp(-(t^2)/2) was not an "accident"--there is a strong theoretical reason why this function is used to model the normalized density function, far stronger than simply that it "looks good on top of the density graph." You'll need to take a statistics course to see why this is true, or at least look through a book on probability and statistics. The result that justifies our use of the exponential is called The Central Limit Theorem.

The value for c is also not accidental. If fact, you can actually figure out what it must be by using the following fact: the integral of the density function from negative Infinity to positive Infinity must equal one (why?!). Hence the value of c must be the reciprocal of the following definite integral value. Compute it! Enter the following command:

```
> evalf(Int(exp(-t^2/2), t=-infinity..infinity));
> 1/";
```

Finally, we note that Maple does have a built-in function that allows the evaluation of the indefinite integral of exp(-t^2/2). This, in turn, will yield a method for computing F(b)-F(a) which is even more efficient than the numerical integration technique developed in the previous section. Let's investigate. Enter the following indefinite integration command:

```
> int(exp(-t^2/2), t);
```

The odd sounding function erf(x) is called the error function, and it is defined by

erf := x -> 2/sqrt(Pi) * int(exp(-t^2), t=0..x);

```
> ?erf
```

There are routines built into Maple to allow for the efficient evaluation of erf(x). To see how these routines for evaluation of erf(t) compare with ordinary numerical integration, examine the output of the following computations. Enter the following commands:

```
> (2/evalf(sqrt(Pi)))*int(exp(-t^2), t=0.. 1.5);
> erf(1.5);
```

□

LAB 22: The p-Series and Improper Integrals

□ NAMES:

□ PURPOSE

For this lab we need to initialize the Maple procedures in order to proceed.
Enter the following line.

```
> with(lab22);
```

For a fixed value of p>0 we investigate the limiting behavior of sums of the form

```
> Display(1);
```

$$\sum_{k=1}^{n} \frac{1}{k^p} = \frac{1}{1^p} + \frac{1}{2^p} + \frac{1}{3^p} + \ldots + \frac{1}{n^p}$$

as n gets arbitrarily large. These sums are called the p-series. The question we pose is: for what values of p do the p-series have finite limiting values as n increases?

The answer will come from considering the limiting behavior of integrals of the form

```
> Display(2);
```

$$\int_{1}^{n} \frac{1}{x^p}\, dx$$

as n approaches infinity. The limiting values of definite integrals such as these are called improper integrals.

□ PREPARATION

It is helpful, but not essential, that you read Section 11.4, pages 737 - 745, in Calculus: Modeling and Application, and work Exploration Activity 2 in that section.

□ LABORATORY REPORT INSTRUCTIONS

Note: Follow your instructor's instructions if they differ from those given here.

For this laboratory you are not required to submit a formal report. Rather you are asked to submit a copy of this Worksheet with the questions answered in the indicated locations. Only one submission will be made. Write in complete sentences and be sure the sentences connect together in coherent ways.

□ PROJECT

¤ 1. Limits of p-series by Brute Force.

Let's try and determine the limit of the p-series sums

```
> Display(1);
```

$$\sum_{k=1}^{n} \frac{1}{k^p} = \frac{1}{1^p} + \frac{1}{2^p} + \frac{1}{3^p} + \ldots + \frac{1}{n^p}$$

as n grows larger and larger by computing the sums for a lot of n values--maybe we'll "see" an approach to a limiting value.

• The Case p=2.

We'll start with p=2. We want to calculate and examine the following sequence of numbers:

```
> Display(3);
```

$$s_1 = \frac{1}{1^2}$$

$$s_2 = \frac{1}{1^2} + \frac{1}{2^2}$$

$$s_3 = \frac{1}{1^2} + \frac{1}{2^2} + \frac{1}{3^2}$$

$$\vdots$$

$$s_n = \frac{1}{1^2} + \frac{1}{2^2} + \frac{1}{3^2} + \ldots + \frac{1}{n^2}$$

$$\vdots$$

Calculate the first five sums s1, s2, s3, s4, and s5 (use whatever method you desire). Enter your computations below.

```
>
```

Performing a separate calculation for each sn is tedious. Let's be efficient and define a function sTwo(n) whose value is the sum of the numbers 1/k^2 (numerical form) for k going from 1 to n:

sTwo(n) = 1/1^2 + 1/2^2 + 1/3^2 + ... + 1/n^2

Enter the following definition for sTwo(n).

```
> sTwo := proc(n) sum(1.0/k^2,k=1..n); end:
```

We will use a do command to calculate values of sn = sTwo(n) for inspection. For example,

```
> for n from 10 to 25 by 5 do
>   printf(`%4d     %f\n`,n,sTwo(n));
> od:
```

prints n and sTwo(n) for the values n = 10, 15, 20, and 25. Use this command (with appropriate parameters) to attempt to determine if the sums sTwo(n) remain bounded as n gets large. Complete and use the following command for your computations.

```
> for n from ??? to ??? by ??? do
>   printf(`%4d     %f\n`, n, sTwo(n));
> od:
```

Explanation. What do we mean by a sequence of numbers s1, s2, s3, ..., sn, ... remaining bounded as n gets large? Well, consider two examples.

Suppose sn = Log(n). Then sn does not remain bounded as n gets large because no matter how large a number M you pick, we can always find an n such that Log(n)>M (simply pick n>Exp(M)). Hence Log(n) has "no ceiling" for its allowable values, and is thus unbounded.

Suppose, however, that sn = 1 - 1/n. Then sn, while always increasing as n increases, does have a "ceiling" for its values: all are less than 1, no matter how large n becomes. Thus sn is a bounded sequence of numbers.

So which is sTwo(n) like, Log(n) (unbounded) or 1 - 1/n (bounded)?

What do you conclude from your computations? In particular:

(1) Do you think the sums for p=2 (i.e., sTwo(n)) remain bounded as n gets large?

(2) Do you think the sums for p=2 have a finite limiting value as n gets large?

(3) If so, what is your approximation for the limiting value?

(4) How confident are you about your answers to (1), (2), and (3) ... and why? Place your answers below.

Answer:

• The Case p=1/2.

Now we'll turn our attention to another value: p=1/2. We want to calculate and examine the following sequence of numbers:

```
> Display(4);
```

$$s_1 = \frac{1}{1^{1/2}}$$

$$s_2 = \frac{1}{1^{1/2}} + \frac{1}{2^{1/2}}$$

$$s_3 = \frac{1}{1^{1/2}} + \frac{1}{2^{1/2}} + \frac{1}{3^{1/2}}$$

.

.

$$s_n = \frac{1}{1^{1/2}} + \frac{1}{2^{1/2}} + \frac{1}{3^{1/2}} + \ldots + \frac{1}{n^{1/2}}$$

.

.

Calculate the first five sums s1, s2, s3, s4, and s5. Enter your computations below.

Let's be efficient and define a function sHalf(n) whose value is the sum of the numbers 1/k^(1/2) for k going from 1 to n:

sHalf(n) = 1/1^(1/2) + 1/2^(1/2) + 1/3^(1/2) + ... + 1/n^(1/2)

Enter the following definition for sHalf(n).

```
> sHalf := proc(n) option remember;
>  sum(1.0/k^(1/2), k=1..n); end:
```

Attempt to determine by inspection if the sums sn = sHalf(n) remain bounded as n gets large. Complete and use the following command for your computations.

```
> for n from ??? to ??? by ??? do
>  printf(`%4d    %f\n`,n,sHalf(n));
> od:
```

What do you conclude from your computations? In particular:

(1) Do you think the sums for p=1/2 (i.e., sHalf(n)) remain bounded as n gets large?

(2) Do you think the sums for p=1/2 have a finite limiting value as n gets large?

(3) If so, what is your approximation for the limiting value?

(4) How confident are you about your answers to (1), (2), and (3) ... and why? Place your answers below.

Answer:

• The Case p=1.

We turn our attention to our third and last value: p=1. We want to calculate and examine the following sequence of numbers:

```
> Display(5);
```

$$s_1 = \frac{1}{1}$$

$$s_2 = \frac{1}{1} + \frac{1}{2}$$

$$s_3 = \frac{1}{1} + \frac{1}{2} + \frac{1}{3}$$

.

.

$$s_n = \frac{1}{1} + \frac{1}{2} + \frac{1}{3} + \ldots + \frac{1}{n}$$

.

.

The p-series for p=1 is more commonly known as the Harmonic Series. Calculate the first five sums s1, s2, s3, s4, and s5 of the Harmonic Series. Enter your computations below.

```
>
```

Let's be efficient and define a function sOne(n) whose value is the sum of the numbers 1/k (numerical form) for k going from 1 to n:

sOne(n) = 1/1 + 1/2 + 1/3 + ... + 1/n

Enter the following definition for sOne(n).

```
> sOne := proc(n) sum(1.0/k,k=1..n); end:
> sOne(900);
```

Attempt to determine by inspection if the sums sn = sOne(n) remain bounded as n gets large. Enter your computations below.

```
> for n from ??? to ??? by ??? do
>   printf(`%4d    %f\n`,n,sOne(n));
> od:
```

What do you conclude from your computations? In particular:

(1) Do you think the sums for p=1 (i.e., sOne(n)) remain bounded as n gets large?

(2) Do you think the sums for p=1 have a finite limiting value as n gets large?

(3) If so, what is your approximation for the limiting value?

(4) How confident are you about your answers to (1), (2), and (3) ... and why? Place your answers below.

Answer:

• A Comparison of p=1/2, p=1, and p=2.

The command pSeriesSummary produces a table of values for the sums sn with p=1/2, p=1, and p=2. This will allow you to compare the behavior of the p-series

```
> Display(1);
```

$$\sum_{k=1}^{n} \frac{1}{k^p} = \frac{1}{1^p} + \frac{1}{2^p} + \frac{1}{3^p} + \ldots + \frac{1}{n^p}$$

for the three values p=1/2, p=1, and p=2. Obtain information on the form of pSeriesSummary with the usual ? command. Complete, then Enter, the following command.

```
> nmin := ???;  nmax := ???;  nstep := ???;
> pSeriesSummary(nmin,nmax,nstep);
```

On the basis of the table just produced, what might you guess the behavior of the p-series is for p between 1/2 and 1? What about between 1 and 2? How confident are you of your answers? Place your answers below.

Answer:

As you can see, determining the limiting behavior of a p-series is not easy to do by just looking at values of the series. It would be useful to have a "test" which could be applied to a p-series, the result of which would tell us whether or not the series has a finite limiting value. Such a test does exist: the Integral Test. The rest of the lab will be devoted to developing and applying this test.

¤ 2. The Function f(x) = 1/x and the Harmonic Series.

The behavior of the sums of the Harmonic Series (the p-series for p=1),

```
> Display(6);
```

$$\sum_{k=1}^{n} \frac{1}{k} = \frac{1}{1} + \frac{1}{2} + \frac{1}{3} + \ldots + \frac{1}{n}$$

is intimately connected with the behavior of the following definite integral:

```
> Display(7);
```

$$\int_{1}^{n} \frac{1}{x}\, dx$$

In this section we will explore the relationship between these sums and integrals, especially as n grows large. The objective of our study will be to answer the question "Do the sums of the Harmonic Series have a finite limit as n grows large?"

It is convenient to define the function f(x)= 1/x. Complete and Enter the following definition of f.

```
> f := x -> ???;
```

Generate a special plot of f(x) over the interval 1 <= x <= 4 by using the command IntegralTest. (Use the command ?IntegralTest to find out what the command does and what it requires for input.) Enter your commands below.

```
>
```

Let AreaLower be the combined total area of the blue rectangles,
AreaCurve the area under the curve y=f(x) over the interval 1<=x<=4, and
AreaUpper the combined total area of the blue and pink rectangles.

What is the relationship between these three numbers, i.e., which is the largest? the smallest? Place your answer below.
Answer:

Complete the following equalities:

AreaLower = ?/? + ?/? + ?/? = Sum(?/?, k = ? .. ?)

AreaCurve = Int(?/?, x = ? .. ?)

AreaUpper = ?/? + ?/? + ?/? = Sum(?/?, k = ? .. ?)

• The Key Relationship. The answers to the two previous questions yield a relationship between the two summations and the integral. Write this out below. Complete the following inequality:

Sum(1/k, k = ? .. ?) <= Int(1/x, x = ? .. ?) <= Sum(1/k, k = ? .. ?)

Use the IntegralTest command again, but this time for the interval 1 <= x <= 5. Enter your command below.

```
>
```

How does the key relationship change when n goes from n=4 to n=5? Place the revised relationship below.

• The General Key Relationship. Instead of using n=4 or n=5, take n to be any positive integer. What will be the general key relationship, valid for any integer n? Record the general key relationship below.

Evaluate the definite integral (the middle term of the Key Relationship), either by hand or by using Maple's int command. What is the limit of this definite integral as n approaches infinity?? (Note: The integration command is int(f(x),x=xmin..xmax).) Enter your commands and answers below.
Answer:

```
>
```

• The Main Question. Using the answers to the previous questions, determine whether

```
> Display(8);
```

$$\sum_{k=1}^{n} \frac{1}{k}$$

remains bounded or becomes unbounded as n gets large. Carefully justify your answer! Place your answer below.
Answer:

Having trouble answering this question? If so, then open the following closed cell....

```
> Answer(1);
```

Does the answer to the Main Question surprise you? Does this answer agree with what you guessed in Section 1 for the p=1 case? Do you now think that these sums have a finite limiting value as n gets large? Place your answers below.
Answers:

How quickly do these "partial sums" increase in value? For example, what was the largest sOne(n) value you computed in Section 1? How does the size of this sOne(n) value compare with the size of the corresponding n? Record your observations below.

```
> sOne(2000);
```

You don't get a very large sum with only a few thousand terms. Have any idea as to how many terms need to be added together in order to reach a large sum like, say, 100? Well, it's easy enough to estimate: by the key relationship, the desired sum from 1 to n will differ from log(n) by no more than one (why?). Hence log(n) is a good estimate for the sum; in particular, the sum of the first n terms of the series will exceed 100 when log(n) >= 100. So simply determine that value of n for which log(n) exceeds 100. Enter your computations below.

```
>
```

Assuming Maple takes about 1 second to add up a hundred terms (adjust this estimate for your own machine), estimate how long you would have to wait in order for Maple to add up enough terms to reach a sum greater than 100. (Express your answer in centuries.) Enter your computations below.

```
>
```

Determine whether or not the following statement is True or False (and give a careful justification for your answer, using the example of the harmonic series): "We can experimentally determine if a p-series tends to a finite or infinite limit by simply adding up lots of terms on a computer and visually observing whether the resulting numbers remain bounded or unbounded." Place your answer (and justification) below.

Answer:

¤ 3. The Function f(x) = 1/x^2.

The behavior of the sums of the p-series for p=2,

```
> Display(9);
```

$$\sum_{k=1}^{n} \frac{1}{k^2} = \frac{1}{1^2} + \frac{1}{2^2} + \frac{1}{3^2} + \ldots + \frac{1}{n^2}$$

is intimately connected with the behavior of the following definite integral:

```
> Display(10);
```

$$\int_{1}^{n} \frac{1}{x^2}\, dx$$

In this section we will explore the relationship between these sums and integrals, especially as n grows large. The objective of our study will be to answer the question "Do the sums of the p-series for p=2 have a finite limit as n grows large?"

We now define the function f(x) = 1/x^2. Complete and Enter the following definition for f.

```
> f := x -> ???;
```

Generate a special plot of f(x) over the interval 1 <= x <= 4 by using the command IntegralTest. Enter your commands below.

```
> IntegralTest(???);
```

Let AreaLower be the combined total area of the blue rectangles,
AreaCurve the area under the curve y=f(x) over the interval 1<=x<=4, and
AreaUpper the combined total area of the blue and pink rectangles.

What is the relationship between these three numbers, i.e., which is the largest? the smallest? Place your answer below.
Answer:

Complete the following equalities:

AreaLower = ?/?^? + ?/?^? + ?/?^? = Sum(?/?^?, k = ? .. ?)

AreaCurve = Int(?/?^?, x = ? .. ?)

AreaUpper = ?/?^? + ?/?^? + ?/?^? = Sum(?/?^?, k = ? .. ?)

The Key Relationship. The answers to the two previous questions yield a relationship between the two summations and the integral. Write this out below. Complete the following inequality:

Sum(1/k^?, k = ? .. ?) <= Int(1/x^?, x = ? .. ?) <= Sum(1/k^?, k = ? .. ?)

Use the IntegralTest command again, but this time for the interval 1 <= x <= 5. Enter your command below.

```
>
```

How does the key relationship change when n goes from n=4 to n=5? Place the revised relationship below.

• The General Key Relationship. Instead of using n=4 or n=5, take n to be any positive integer. What will be the general key relationship, valid for any integer n? Record the general key relationship below.

Evaluate the definite integral (the middle term), either by hand or using Maple's int command. What is the limit of this definite integral as n approaches infinity? Enter your commands and answers below.
Answer:

```
>
```

• The Main Question. Using the answers to the previous questions, determine whether

```
> Display(11);
```

$$\sum_{k=1}^{n} \frac{1}{k^2}$$

remains bounded or becomes unbounded as n gets large. Carefully justify your answer! Place your answer below.
Answer:

Does the answer to the Main Question surprise you? Does this answer agree with what you guessed in Section 1 for the p=2 case? Do you now think that these sums have a finite limiting value as n gets large? Place your answer below.
Answer:

Estimate the value of the limit with as much accuracy as you can obtain. State how close you believe your estimate is to the "real" answer. (Note: the value of the integral is not a good approximation for the limit of the sums.) Enter your computations below.

```
>
```

¤ 4. The Function f(x) = 1/x^p and the general p-Series. (Optional--Ask Instructor)

In this section you will determine the range of p values such that the p-Series

```
> Display(12);
```

$$\sum_{k=1}^{n} \frac{1}{k^p}$$

approaches a finite limiting value as n approaches infinity. As in Section 2-3, the procedure is to compare this sum with the definite integral

```
> Display(13);
```

$$\int_{1}^{n} \frac{1}{x^p}\, dx$$

For a fixed p>0, what is the Key Relationship that exists between the sums and the integral? Place your relationship below.

On the basis of this relationship, complete the following statement:

The sums of a p-series will remain bounded as n becomes large (and hence will have a finite limiting value as n increases) if and only if the definite integral... ???

• The Main Question. Using the answers to the previous questions, determine those values of p for which the sum

```
> Display(12);
```

$$\sum_{k=1}^{n} \frac{1}{k^p}$$

approaches a finite or infinite limit as n gets large. Carefully justify your answer! Place your answer below.
Answer:

Compare your conclusions for the general p-series with the guesses you made at the end of Section 1--how good were your guesses? Place your answer below.
Answer:
□

LAB 23: Periodic Functions

□ NAMES:

□ PURPOSE
Many phenomena in the natural and social sciences are well approximated by periodic models, i.e., functions which repeat themselves in a regular pattern. The most basic periodic functions are the sine and cosine. In this lab we consider the construction of other simple periodic functions from the addition of sines and cosines. This construction is relatively easy and not surprising. However, what is both surprising and profound is that, in a certain sense, all periodic functions come from additions of sines and cosines. Moreover, knowing how to decompose a periodic model into sines and cosines gives you extremely important information about the quantity you are modeling.

Therefore, the major goal of this lab is to explore the following question: given a periodic function, how can we express it as an addition of sine and cosine functions? After developing the tools we need--essentially the concept of Fourier Series--we will show how these techniques apply to determining if periodicity is present in a data set.

□ PREPARATION
Read the beginning of Section 9.3, pages 534 - 538, and do Checkpoint 1, 2, Exploration Activities 1, 2, 3, and 4.

□ LABORATORY REPORT INSTRUCTIONS
For this laboratory you are not required to submit a formal report. Rather you are asked to submit a copy of this Worksheet with the questions answered in the indicated locations. Only one submission will be made. Write in complete sentences and be sure the sentences connect together in coherent ways.

□ Maple Procedures. For this lab we need two entries in the next region.

```
> with(lab23);
> randomize():
> # If an Error is the output of the above region, see your lab instructor.
```

□ PRELMINARIES

¤ 1. More on Definite Integrals and Maple: Generic Solutions.

Maple's int command can be applied to many classes of functions that contain unknown constants--this will be important in our subsequent computations. However, to use this feature correctly we need to be aware that the int command yields generic solutions which are not necessarily valid for all values of the unknown constants. Here is an example.

Use Maple to integrate the function x^n. Place your computation below.

```
> int(???, ???);
```

Is the answer you have obtained valid for all values of n? (Hint: no!) For what value of n is the answer inaccurate? n=??? . Giving n this value, recompute the integral using Maple. Place your computation below.

>

Why did Maple make an "error" with this value of n in the first computation? Well, it's not really an "error": the Integrate command simply produces generic solutions when unknown constants are present, i.e., solutions that are given without conditions on the unknown constants. A user is expected to recognize when a generic solution doesn't make sense for a particular value of a constant; if a solution in such an exceptional case is needed, the user must enter the appropriate "exceptional" values and perform a separate integration (as we did in our example above).

In the next section you will generate generic solutions for some definite integrals; it will be important that you recognize the existence of exceptional cases.

¤ 2. Some Definite Integrals.

The definite integrals whose values you will need to justify the computations in this lab are the following:

```
> Display(1);
```

$$SnSm = \int_{-\pi}^{\pi} \sin(n\ t)\ \sin(m\ t)\,dt,\ CnCm = \int_{-\pi}^{\pi} \cos(n\ t)\ \cos(m\ t)\,dt$$

$$SnCm = \int_{-\pi}^{\pi} \sin(n\ t)\ \cos(m\ t)\,dt$$

$$S2n = \int_{-\pi}^{\pi} \sin(n\ t)^2\,dt,\ C2n = \int_{-\pi}^{\pi} \cos(n\ t)^2\,dt$$

Here n and m are non-negative integers. Moreover, for the integrals on the first and second lines (SnSm and CnCm), we further assume that n <> m (or n not equal to m), and for the integrals on the fourth line (S2n and C2n) we assume that n is positive. (If n=m, then SnSm and CnCm become what other integrals on our list? Fill in answers and check them below.

Answer: SnSn = ??? , CnCn = ??? .)

```
> Answer(1);
```

• SnSm and CnCm, n<>m.

```
> Display(2);
```

$$\text{SnSm} = \int_{-\pi}^{\pi} \sin(n\ t)\ \sin(m\ t)\ dt,\ \text{CnCm} = \int_{-\pi}^{\pi} \cos(n\ t)\ \cos(m\ t)\ dt$$

Plot the function sin(2*t)*sin(t) over the interval [-Pi,Pi], and then--without any computations--estimate the value of the definite integral S2S1. Justify your answer. Enter your computations below.

Repeat the previous question for cos(t)*cos(3*t) and the definite integral C1C3. Enter your computations below.

```
>
```

Check your estimate for C1C3 by performing the integration using Maple. Does the result of Maple's integration confirm your estimate? Enter your computations below.

```
>
```

On the basis of the two observations made thus far, are you willing to make conjectures about the integrals SnSm and CnCm for all values of n <> m? If so, what are your conjectures? Place your conjectures below.

Answer: For n <> m we conjecture that SnSm = ??? and CnCm = ??? .

Calculate another of these integrals as a check on your conjecture. Enter your computations below.

```
>
```

Maple is powerful enough to check your conjecture for you--the Integrate command can be used on the definite integrals given above, even with n and m left as unspecified constants. Pick one of the two integrals and attempt to evaluate it using int. Enter your computations below.

- Maple Tip. A common mistake in this lab is to type nt rather than n*t.

```
> int( ??? );
```

Most likely you did not get what you expected! However, examine your answer: what is the value of sin(k*Pi) for any integer k? Does this give you the answer you expected? And why is this computation not valid for n=m? Place your answer below.

- SnCm.

```
> Display(3);
```

$$\text{SnCm} = \int_{-\pi}^{\pi} \sin(n\ t)\ \cos(m\ t)\ dt$$

As done for SnSm and CnCm, use Maple to compute SnCm with n and m left as unspecified constants. Enter your computations below.

```
>
```

You should be pleased with the simplicity of this answer!

• S2n and C2n, n>0

```
> Display(4);
```

$$S2n = \int_{-\pi}^{\pi} \sin(n\ t)^2\, dt, \text{ and }, C2n = \int_{-\pi}^{\pi} \cos(n\ t)^2\, dt$$

Plot the function sin(2*t)^2 over the interval [-Pi,Pi], and then estimate the value of the definite integral S22. Justify your answer. Enter your computations below.

```
>
```

Check your estimates for S2^2 by performing the integration using Maple. Does the result of Maple's integration confirm your estimate? Enter your computations below.

```
>
```

Use Maple to compute all of the integrals S2n, with n any positive integer. Enter your computations below.

```
>
```

Examine your answer: what is the value of sin(k*Pi) for any integer k? What final answer does this give you? And why is the computation not valid for n=0? Place your answer below.

Repeat your computations for C2n. Enter your computations below.

```
> int(C2(n),t=-Pi..Pi);
```

¤ Summary.

Recall the integrals we have considered in this section:

```
> Display(1);
```

$$SnSm = \int_{-\pi}^{\pi} \sin(n\ t)\ \sin(m\ t)\, dt, CnCm = \int_{-\pi}^{\pi} \cos(n\ t)\ \cos(m\ t)\, dt$$

$$SnCm = \int_{-\pi}^{\pi} \sin(n\ t)\ \cos(m\ t)\, dt$$

$$S2n = \int_{-\pi}^{\pi} \sin(n\ t)^2\, dt, C2n = \int_{-\pi}^{\pi} \cos(n\ t)^2\, dt$$

Record all the values you have found for these integrals. Place your answers below.

Answer: If n <> m, then SnSm = ??? and CnCm = ??? .
SnCm = ??? .
If n>0, then S2n = ??? and C2n = ???.

□ PROJECT

¤ 1. Constructing Periodic Functions.

Quite complex periodic functions can be built up out of the "elementary" periodic functions:
sin(t), sin(2*t), sin(3*t),..., and 1, cos(t), cos(2*t), cos(3*t),....
Let's consider some examples.

First we determine the geometric effect of multiplying the independent variable t by an integer. (For example, how does the graph of sin(2*t) differ from the graph of sin(t)?) Plot the following three functions: sin(t), sin(2*t), and sin(3*t) over the interval [0, 4*Pi]. Enter your plots below.

```
>
```

From what you have seen in these three plots, describe what happens to the period and the frequency of the function sin(m*t) as m increases through the integer values m=1, 2, 3, ...Enter your answer below.

What do you think happens to the functions cos(m*t) as m increases through the integer values m=0, 1, 2, 3, ...? Enter your answer below.

Now let's consider what results from forming linear combinations of the "elementary" periodic functions. In particular, we are interested in seeing what happens when functions with different frequencies are added together.

Execute the following Plot statements, one at a time:

```
> plot( 1+2*cos(t), t = 0 .. 6*Pi );
```

• Comment: The graph of 1+2cos(t) is just a simple cosine function translated vertically.

```
> plot( 1+2*cos(t)+4*sin(t), t = 0 .. 6*Pi );
```

• Comment: Adding the term 4*sin(t) still yields a graph with a simple sine form, but now translated horizontally so that the center is no longer at the origin.

```
> plot( 1+2*cos(t)+4*sin(t)-6*sin(2*t), t = 0 .. 6*Pi );
```

• Comment: The subtraction of the term 6*sin(2*t), whose frequency is double that of the original terms, makes the resulting graph considerably more interesting! See what happens when you add in another term of even higher frequency (this time you choose the term to add on).

Complete and Execute the following Plot statement.

```
> plot( 1+2*cos(t)+4*sin(t)-6*sin(2*t) + ???, t = 0 .. 6*Pi);
```

As you can see, this process of constructing periodic functions from linear combinations of the "elementary" trigonometric functions can yield surprisingly complex results.

¤ 2. Decomposing Periodic Functions.

When you initialized this Maple Worksheet, you defined a function Func(t) of the form

Func(t) = b0 + b1*cos(t) + b2*cos(2*t) + a1*sin(t) + a2*sin(2*t).

The five constants b0, b1, b2, a1, and a2 all have been given specific values. The purpose of this section is to develop the techniques necessary to determine the values for b0, b1, b2, a1, and a2.

• Caution: To keep the lab interesting, the code for Func(t) has been written so that every time you initialize the Worksheet, a different set of values for the constants b0, b1, b2, a1, and a2 will be used. In particular, no two laboratory groups will have the same Func, and if you reinitialize the Worksheet, your function Func will be changed.

Graph Func(t) over the interval [0, 4*Pi]. Enter your Plot below.

```
>
```

Your task is to use integration to find the values of the constants b0, b1, b2, a1, and a2. In fact, we claim that each of these constants can be expressed as a definite integral of Func multiplied by a suitable elementary trigonometric function. To obtain the desired formulas we proceed as follows:

• The First Integral Formula.

We intend to compute the value of the following integral.

```
> Display(5);
```

$$\int_{-\pi}^{\pi} F(t)\, dt$$

where F is in the general form specified at the beginning of this section, i.e.,

F(t) = b0 + b1 cos(t) + a1*sin(t) + b2*cos(2*t) + a2*sin(2*t)
but with the constants b0, b1, b2, a1, and a2 left as unknown letter symbols. In order to do this we define the function F(t) with the following command. Enter the following command.

```
> F := t -> b0 + b1*cos(t) + a1*sin(t) + b2*cos(2*t) + a2*sin(2*t);
```

Notice that the function Func(t) as defined at the beginning of the section is simply a particular example of F(t), with the constants given some specific values.

Use Maple to compute the definite integral of F(t) which we displayed above. Enter your computations below.

```
>
```

The computation you just did yields "an equation," i.e.,
The definite integral of F(t) from -Pi to Pi = a (simple) expression.
This can be turned around to give an integral formula for one of the unknown constants b0, b1, b2, a1, or a2. Do so. Complete the following integral formula.

Answer: ??? * (1/(???))

```
> int(???, t = ??? .. ???)/(???);
```

Question: How could you have computed the integral of F(t) without Maple?
Answer: By observing that the integrals of cos(t), sin(t), cos(2*t), and sin(2*t) are all zero over the interval from -Pi to Pi. This will be examined in class.

- The Second Integral Formula.

Now use Maple to compute the following integral.

```
> Display(6);
```

$$\int_{-\pi}^{\pi} F(t)\, \cos(t)\, dt$$

Enter your computations below.

```
>
```

Turn this equation around to give an integral formula for another of the unknown constants b0, b1, b2, a1, or a2. Complete the following integral formula.

Answer: ??? * (1/ ???)

```
> int( ??? , t = ??? .. ???)/???;
```

• All the Integral Formulas.

Continue this process, determining integral formulas for all five of the unknown coefficients b0, b1, b2, a1, and a2. Place your results below.

b0 = (1/ ???) * int(???, t = ??? .. ???);
b1 = (1/ ???) * int(???, t = ??? .. ???);
b2 = (1/ ???) * int(???, t = ??? .. ???);
a1 = (1/ ???) * int(???, t = ??? .. ???);
a2 = (1/ ???) * int(???, t = ??? .. ???);

You now need to apply these formulas to the specific function F(t) = Func(t). To do so, rewrite the formulas with F(t) replaced by Func(t). Complete the following formulas.

```
> b0 := evalf(int(???,t=???..???)/(???));
> b1 := evalf(int(???,t=???..???)/???);
> b2 := evalf(int(???,t=???..???)/???);
> a1 := evalf(int(???,t=???..???)/???);
> a2 := evalf(int(???,t=???..???)/???);
>
```

Entering these formulas will compute the values of b0, b1, b2, a1, and a2 for the specific function Func defined in the Maple Packages for this lab.

If you have not already done so, Enter the above formulas.

Note: How could you have computed the integrals of F(t)*cos(t), F(t)*sin(t), etc., without Maple? Answer: By using the integral formulas for SnSm, CnCm, SnCm, S2n, and C2n as determined in the Preliminaries section. This will be examined further in class.

• The Mystery Function Exposed.

The code below takes the just-computed values for b0, b1, b2, a1, and a2 and defines a new function FuncExposed:

if your formulas for b0, b1, b2, a1, and a2 are correct, FuncExposed will equal Func.

We check this equality graphically by plotting Func and FuncExposed on the same set of coordinate axes. Enter the following lines which define, display, and plot FuncExposed.

```
> FuncExposed(t) := b0 + b1*cos(t) +b2*cos(2*t) + a1*sin(t) +a2*sin(2*t):
> FuncExposed(t);
> plots[display]([ plot(Func(t),t=0..4*Pi,color=black,thickness=2,title=`FuncExposed=yellow,
> Func = black`), plot(FuncExposed(t),t=0..4*Pi,color=yellow,thickness=2) ]);
```

Are the two curves in your plot superimposed on each other? If not, you had better go back and check your formulas for the constants b0, b1, b2, a1, and a2.

¤ 3. Fourier Series.

Suppose we have a function f(t) with more than five trigonometric terms; for example,

f(t) = b0 + b1*cos(t) + b2*cos(2*t) + ... + bn*cos(n*t)
+ a1*sin(t) + a2*sin(2*t) + ... + an*sin(n*t)

for some integer n. The question is: can we find integral formulas for the general k-th coefficients ak and bk?

If you examine the five integral formulas derived in the previous section, you should see a pattern that strongly suggests the formulas for ak and bk. Place your conjectured formulas below, then check against the next cell.

Answer: ak = (1/ ???) * int(???, t = -Pi .. Pi), bk = (1/ ???) * int(???, t = -Pi .. Pi)

```
> Answer(2);
```

These formulas for ak and bk are proven with the same techniques that you used to obtain your five integral formulas for b0, b1, b2, a1, and a2, i.e., by using the integral formulas determined in the Preliminaries Section.

Notice that the integrals in the formulas for ak and bk can be computed for essentially any function f(t) which is defined on the interval [-Pi,Pi], i.e., f(t) does not have to be an addition of sines and cosines in order for us to define the numbers ak and bk. This motivates the following:

Given f(t) on [-Pi,Pi], the 0-th order Fourier coefficient for f(t) is the number b0 defined by

```
> Display(7);
```

$$b_0 = \frac{1}{2}\,\frac{\int_{-\pi}^{\pi} f(t)\,dt}{\pi}$$

while the k-th order Fourier coefficients for f(t) (k>0) are the numbers ak and bk defined by

```
> Display(8);
```

$$a_k = \frac{\int_{-\pi}^{\pi} f(t)\sin(k\,t)\,dt}{\pi}, \text{ and }, b_k = \frac{\int_{-\pi}^{\pi} f(t)\cos(k\,t)\,dt}{\pi}$$

The n-th order Fourier expansion fn(t) for f(t) is the sum of the Fourier coefficients of degree less than or equal to n, each times the appropriate sine or cosine function, i.e.,

fn(t) = b0 + b1*cos(t) + b2*cos(2*t) + ... + bn*cos(n*t)
+ a1*sin(t) + a2*sin(2*t) + ... + an*sin(n*t).

As an example, compute the 5-th order Fourier coefficients for f(t)=t-t^2. Enter your computations below.

```
> a5 := ???;
> b5 := ???;
```

Of what significance are the Fourier expansions fn(t) for a function f(t)? Let's consider this question with an example. Pick a function f(t) and we'll examine the Fourier expansions for f(t). Enter your definition for f(t) below.

```
> f := t -> ???
> FourierSeries(f,0) := evalf(b(f,0)/2):
> # This is needed to initialize the recursive definition of
> # FourierSeries
```

Caution: Be sure your function f(t) makes sense for all t in the interval [-Pi,Pi].

We have placed the Fourier expansion formulas into a command called FourierSeries. Determine information about this command in the usual way. Enter your command below.

```
> ?FourierSeries
```

Use FourierSeries to compute at least the 6-th order Fourier expansion for your function f(t).

Enter your command below.

```
>
```

Animate the process by the FourierSeriesMovie command and then describe the behavior of the k-th Fourier expansion as k increases. Place your answer below.

```
> FourierSeriesMovie(f,0,6);
```

¤ 4. A Finite Fourier Series for Data. (Somewhat sketchy.)

Suppose we have n data points [tk, qk], k=1,...,n, for a t dependent quantity Q(t), i.e., qk=Q(tk) for each k. Our concern is: how do we detect periodic behavior in this data? More specifically, how can we analyze this data in order to detect fundamental periodic behavior in the quantity Q?

To simplify the discussion, let's assume that the points tk, k=0,...,n are evenly spaced in the interval [-Pi,Pi] (Thus t0=-Pi. We will not, however, use a measurement of Q at t0.) In this case we can fit our data with a finite Fourier series of the following form:

```
> Display(9);
```

$$q_k = b_0 + b_1 \cos(t_k) + b_2 \cos(2\, t_k) + \ldots + b_{n-1} \cos((n-1)\, t_k) + a_1 \sin(t_k) + a_2 \sin(2\, t_k) + \ldots + a_{n-1} \sin((n-1)\, t_k)$$

where the coefficients ak and bk are given by the formulas

```
> Display(10);
```

$$b_j = \frac{\sum_{k=1}^{n} q_k \cos(t_k\, j)}{n}, \quad j = 0, \ldots, n-1$$

$$a_j = \frac{\sum_{k=1}^{n} q_k \sin(t_k\, j)}{n}, \quad j = 1, \ldots, n-1$$

(We will omit the calculations that lead to these formulas for aj and bj. Note, however, that they resemble the formulas you had earlier for the Fourier coefficients ak and bk of a function y=f(t). The major difference is that the integrals are replaced by summations.)
Here is our important claim:

- The quantity Q has some periodicity with frequency k if one or both of the k-th Fourier coefficients, ak and bk, are much larger in magnitude than most of the other coefficients.

Here is a data set to be analyzed for periodicity. First plot it, then apply the command DiscreteFourierCoefficients to analyze the frequencies. Identity as many frequencies as seem to be significant in the output--does this data have any significant periodicity? Enter your commands and answers below.

```
> data := [[1, -2.087218149803781677], [2, -1.626903151713628632], [3, -0.937523125018383
> 4436], [4, 0.394958638291514340 2], [5, 1.038336531730868238], [6, 1.27456557048985386
> 6], [7, 1.097862086735942299], [8, 0.857747786639622891], [9, -1.185984411803787 49], [1
> 0, -1.396542120741195098], [11, -2.129502816195029399], [12, -2.357072469168346131], [
> 13, -1.201766305275068368], [14, 0.300058340756417286 9], [15, 1.578661846128375159], [
> 16, 2.388469165059521 66], [17, 2.786323054822156392], [18, 1.143047637412609244], [1
> 9, 0.422145448792066419 6], [20, -0.256315467678353503 1], [21, -0.5644674605766207464]
> , [22, -0.181474015250987154], [23, 1.150714587817674911], [24, 2.582790305577695 31],
> [25, 3.315962630648554669], [26, 3.242000758375073755], [27, 3.087942943819977413], [2
> 8, 1.390702045380081697], [29, -0.5072110912804762233], [30, -0.9295821318908930711],
>  [31, -1.616161205433794729], [32, -0.6376279908539691402]];
> FourierCoeffs := ???;
```

• This final graph shows the frequency k along the independent axis and the magnitude of the Fourier coefficients ak and bk along the dependent axis. Identity as many frequencies as seem to be significant in the output. Which are the most important frequencies? Does this data have any significant periodicity? Place your answers below.

We can plot the portions of the trigonometric expansion which exhibit the "significant periodicity" by using the discrete Fourier coefficients as returned by DiscreteFourierCoefficients. Choose a range of frequencies [kMin,kMax] (where 1<=kMin<=kMax<=nData/2), and then run the following set of commands. If you have picked up the "significant periodicities," then the red curve you obtain should approximate the scatter plot of the data. Complete and Enter the following commands.

```
> kMin := ???:
> kMax := ???:
> b[0] := FourierCoeffs[1]:
> for k from 1 to nData - 1 do
>   a[k] := FourierCoeffs[k+1][1];
>   b[k] := FourierCoeffs[k+1][2];
> od:
> TrigApprox := (b[0] + 2*sum(a[l]*sin(l*t)+b[l]*cos(l*t),l = kMin .. kMax) ):
> trigPlot := plot(TrigApprox,t = -1.1*Pi .. 1.1*Pi,color=red):
> plots[display]([dataPlot,trigPlot]);
```

Did your plot confirm your claims as to the "significant periodicities" in the data? Place your answer below.

□

LAB 24: Taylor Polynomials I

☐ NAMES:

☐ PURPOSE
Recall: In a previous lab we used definite integration to determine the coefficients b0, b1, b2, a1, and a2 of a function of the form

f(t) = b0 + b1*cos(t) + b2*cos(2*t) + a1*sin(t) + a2*sin(2*t) .

We then used this information to approximate an arbitrary periodic function by "trigonometric polynomials," i.e., by linear combinations of sin(n*t) and cos(m*t), for n and m non-negative integers.

New: In this lab we use differentiation to determine the coefficients a0, a1, a2, ..., an-1, an of any polynomial of the form

p(x) = a0 + a1*x + a2*x^2 + a3*x^3 + ... + an*x^n

We will then use this information to approximate an arbitrary function by polynomials--the Taylor polynomials for the function under consideration.

☐ PREPARATION
Study Sections 11.1 and 11.2, pages 701 - 720, in Calculus: Modeling and Application.

☐ LABORATORY REPORT INSTRUCTIONS
Note: Follow your instructor's instructions if they differ from those given here.

For this laboratory you are not required to submit a formal report. Rather you are asked to submit a copy of this worksheet with the questions answered in the indicated locations. Only one submission will be made. Write in complete sentences and be sure the sentences connect together in coherent ways.

☐ Maple Procedures. Activate the following command region to have access to the procedures written for this lab.

```
> with(lab24);
> randomize():
> # If an Error is the output of the above line, see your lab instructor.
```

The first procedure for this lab, RandomUnknownPolynomial, defines a polynomial Poly and its derivatives DPoly for use in the project. Each time the package is evaluated a different polynomial will be constructed (the coefficients are chosen via the Random function).

□ PRELIMINARIES

¤ Derivation of the Formulas for the Taylor Coefficients--A Review.

Suppose we know a polynomial of the form

p(x) = a0 + a1*x + a2*x^2 + a3*x^3 + ... + an*x^n

in the sense that we can obtain values of the polynomial and its various derivatives at x=0. We would like to find the values of the coefficients a0, a1, a2, ..., an.

Obtain a formula for a0 in terms of p. (Hint: What is p(x) at x=0?) Write your formula on paper, then Enter the following command.

a0 = ???

```
> Answer(1);
```

Calculate the derivative of p(x) and evaluate at x=0. A formula for a1 in terms of p can be obtained from your formula for p'(0). Write your formula on paper, then Enter the following command.

p'(0) = ???   a1 = ???

```
> Answer(2);
```

Calculate the second derivative of p(x) and evaluate at x=0. This yields a formula for a2. Write your formula on paper, then Enter the following command.

p''(0) = ???   a2 = ???

```
> Answer(3);
```

Calculate the third derivative of p(x) and evaluate at x=0. This yields a formula for a3. Write your formula on paper, then Enter the following command.

p'''(0) = ???   a3 = ???

```
> Answer(4);
```

Find a general expression for ak, k any non-zero integer, and check it by calculating a4. (Helpful notation: the product of the first k positive integers is called k factorial and is denoted by k!. Thus, for example,

1! = 1
2! = 1*2 = 2
3! = 1*2*3 = 6
4! = 1*2*3*4 = 24
5! = 1*2*3*4*5 = 120.

Although it may seem strange, we also define 0! = 1. This definition for 0! fits well into the formulas in which it is used.) Write your formula on paper, then Enter the following command.

p(k)(0) = ??? ak = ???

```
> Answer(5);
```

□ PROJECT.

¤ 1. Determining the Coefficients of a Polynomial.

When you initialized this worksheet a polynomial Poly(x) was constructed of the form

Poly(x) = a0 + a1*x + a2*x^2 + a3*x^3 + ... + an*x^n

The coefficients a0, a1, a2, ..., an have all been given specific values. The purpose of this section is to determine the values of the coefficients by using the Taylor polynomial formulas.

• Note: To keep the lab interesting, the code for Poly(x) has been written so that every time you initialize the worksheet, a different set of values for the coefficients a0, a1,..., an will be used. In particular, if you reinitialize the worksheet, your function Poly will be changed.

Let's begin our analysis by plotting the polynomial. In the following Plot command vary the x limits until you can see all the important portions of the graph of Poly. Start with the limits xmin = -10 and xmax = 10 then, if necessary, change them to get the best picture. Complete and Enter your command below as often as necessary.

```
> xmin := ???: xmax := ???:
> fPlot := plot(Poly(x),x=xmin..xmax):";
```

You will use the values of Poly(x) and its derivatives to calculate the a(0), a(1), a(2), etc.

• The n-th derivative of Poly at x=x0 is given by the command DPoly(n, x0).

As an example, use this command to compute the second derivative of Poly at x=1.23. Complete and Enter your command below.

```
> DPoly(???, ???);
```

In the Preliminaries section you determined a formula for the coefficients of a polynomial in terms of its derivatives. Place this formula for a(k) into the following for statement in order to compute the coefficients for Poly. Complete and Enter the following command.

• Remember: You've been given a special command DPoly(n,x) to compute the derivatives of Poly(x). You need to use this command in your formula for a(k). Also, Maple does know the meaning of the factorial symbol k!.

```
> for k from 0 to 10 do
>   a[k] := ???;
>   printf(`a[%2d] = %12.6f\n`,k,a[k]);
> od:
```

• Maple Tip. The output of this last command should be numerical values for the coefficients a(0), a(1), a(2), ... Do not go on until you obtain numerical values for these quantities.

We now use these coefficients to define the Taylor polynomials for Poly(x), i.e., for each n=1,2,3,...,10 we define the polynomial p(n,x) to be the sum of all the terms a(k)*x^k for k between the values of 0 and n:

p(n,x) = a0 + a1*x + a2*x^2 + a3*x^3 + ... + an*x^n

Enter the following command to define p(n,x).

```
> p := proc(n,x) local k; a[0] + sum(a[k]*x^k, k=1..n); end:
```

Display the 10th Taylor polynomial p(10,x). Does it look correct? Enter the following command.

```
> p(10,x);
```

We now plot each of the functions p(1,x), p(2,x), ..., on top of the graph of Poly. This is easily carried out by the new command TaylorPlot. Complete and Enter the following command.

```
> ymin := ???: ymax := ???:
> TaylorPlot([Poly(x),p(n,x)],x=xmin..xmax,  n=0..10,1,y=ymin..ymax);
```

Animate the plots produced above, studying the relationship between the graphs of the polynomials p(1,x), p(2,x),..., and the graph of Poly(x). Perform your animation above.

What is the relationship between the graphs of p(1,x), p(2,x),..., and the graph of Poly(x)? Is there an n for which p(n,x)=Poly(x)? Place your answers below.

Answer:

What is Poly(x)? (Note: to show Poly(x), simply Enter the next expression.

```
> Poly(x);
```

¤ 2. Taylor Polynomials for the Exponential Function.

We will now consider a function that is not a polynomial, the exponential function f(x) = e^x. In proper Maple notation this will be denoted as f(x)=exp(x). We start by plotting the function--for our purposes a good interval to plot over is xmin=-6 to xmax=4. Plot the exponential curve below.

```
> f := x -> exp(x);
> xmin := -6: xmax := 4:
> plot ???
```

Exp(x) is not a polynomial, and hence our original definition of the numbers a(0), a(1), ..., as the coefficients for the xk terms in the "polynomial" exp(x) makes no sense. However...,

• the derivative formulas found in the Preliminaries for the a(k) still "make sense" for non-polynomial functions such as exp(x) (i.e., these formulas can be evaluated whether or not the function in question is a polynomial).

As we will attempt to make clear in this lab, the resulting numbers a(0), a(1), a(2), ... can be used as coefficients of polynomials which approximate exp(x). These will be the Taylor Polynomials for exp(x).

Using the formulas for the a(k), derived in the Preliminaries section, compute the a(k) for exp(x). (Do this by hand, not with Maple.) Place your (numerical) results below.

a(0) = ???
a(1) = ???
a(2) = ???
a(3) = ???
a(4) = ???
In general,
a(k) = ???

Place your (simple!) formula for a(k) into the following Do statement in order to compute the coefficients. Complete and Enter the following command.

```
> for k from 0 to 12 do
>   a[k] := ???;
>   printf(`a[%2d] = %7.10f\n`,k,a[k]);
> od:
```

For each n=1,2,3,...,12 define the polynomial p(n,x) to be the sum of all the terms ak*x^k which are of degree <= n, i.e.,

p(x) = a0 + a1*x + a2*x^2 + a3*x^3 + ... + an*x^n

For n sufficiently large do you think that p(n,x) = exp(x)? You will now test your claim. Enter the following command in order to define p(n,x).

```
> p := proc(n,x) local k; a[0] + sum(a[k]*x^k,k=1..n) end:
```

Display the 10th Taylor polynomial p(10,x). Does it look correct? Enter the following command.

```
> p(10,x);
```

Now plot each of the functions p(1,x), p(2,x), ..., on top of the graph of exp(x). This is carried out by the command TaylorPlot. Complete and Enter the following command.

```
> TaylorPlot([exp(x), p(n,x)], x = xmin .. xmax, n = 0 .. 10, 1);
```

What is the relationship between the graphs of p(1,x), p(2,x),..., and the graph of exp(x)?. Is there an n for which p(n,x)=exp(x)? Place your answers below.

Answer:

If you conjectured the existence of an n for which p(n,x)=exp(x), then be sure to "double check" your claim by increasing the size of the x-interval in TaylorPlot (i.e., replace xmin and xmax with a wider interval, say -10 to 10) and executing the command again. Does your claimed equality still remain valid? Place your answers below.

Answer:

Display the highest degree Taylor polynomial p(n,x) that you computed for exp(x). Complete and Enter your command below.

```
> p(???,x);
```

¤ 3. The Error Terms.

We continue to work with the polynomials p(n,x) defined for the exponential function.

• Be sure that the definition of p(n,x) has been Entered in the previous section, along with the coefficients a(0), a(1), a(2),....

In order to measure how good our polynomial approximations for exp(x) really are, we plot the approximation errors for the various polynomials p(n,x). These error terms are merely the differences between exp(x) and each polynomial p(n,x), i.e.,

errorn = exp(x) - p(n,x).

Hence the first few error terms are:

error1 = exp(x) - 1 - x
error2 = exp(x) - 1 - x - (x^2)/2
error3 = exp(x) - 1 - x - (x^2)/2 - (x^3)/6

If p(n,x) is a good approximation of exp(x), then the absolute value of exp(x)-p(n,x) is small. If p(n,x) is a bad approximation of exp(x), then the absolute value of exp(x)-p(n,x) is large.

We will plot all of the error terms on one graph using the new command TaylorErrorPlot. Enter the following command.

```
> TaylorErrorPlot([exp(x),p(n,x)], x = -3.1 .. 2.1, n = 1 .. 12, 1,y=-1..2.5);
```

The bright red "parabola-like" curve in the center of the plot is the graph of the n=1 error term,

y = error1 = exp(x) - p(1,x) = exp(x) - 1 - x

The deeper red "cubic-like" curve which is next is the graph of the n=2 error term,

y = error2 = exp(x) - p(2,x) = exp(x) - 1 - x - x^2

The error curves can be used to approximate values of exp(x) to specified accuracies. For example, to approximate exp(5/4) to within an error of at most 0.01, use the plot above to determine the smallest value of n such that |errorn| at x=5/4 is less than 0.01. Then p(n,5/4) is the desired approximation to exp(5/4), i.e., p(n,5/4) differs from exp(5/4) by no more than 0.01. Determine this value of n and evaluate the corresponding approximation p(n,5/4). Check your answer against the "actual" value of exp(5/4). Enter your computations and explanations below.

Explanations:

>

A thought question. We claim that the use of the error curves in the previous computation is highly artificial, i.e., if we really needed to approximate values of the exponential function, the error curves would not be available to us. Why would these curves not be available? (Hint: What do you need to compute in order to construct the error curves?) Place your answer below.

Answer:

Although we cannot compute the error curves in those situations where we wish to use the Taylor polynomials to estimate function values, there is a formula for estimating the size of the errors: the Taylor Remainder Formula. This will be discussed later in the semester.

¤ 4. Taylor Polynomials for the Cosine Function.

The polynomials described in Section 2 and 3 are called the Taylor polynomials for the function exp(x). Maple can automatically compute most Taylor polynomials in the following way. Enter the command given below.

```
> series(exp(x),x=0,11);
```

This should be the 10th degree polynomial for the exponential function that you obtained in the previous section. The O(x^11) at the end of the polynomial is the "remainder". This term is added in to insure that the polynomial expansion is actually equal to the function handed in.

Hence, given any "reasonable" function f(x), we can compute its n-th Taylor Polynomial by the following command:

```
series(f(x), x=0,n);
```

For convenience we embed this statement into a new command, TaylorPoly, that is configured to produce a well-behaved polynomial function of n and x. Obtain information about TaylorPoly by using the ? command, and then use TaylorPoly to compute the tenth Taylor polynomial for the function cos(x). Complete and Enter the commands below.

```
> ?TaylorPoly
```

```
> TaylorPoly(???);
```

What are the most striking features of the Taylor polynomial you've just produced? Place your answer below.

Answer:

The graphing command TaylorPlot, introduced in the previous sections, has a useful feature: when given appropriate parameters, it will itself calculate the Taylor polynomials for the given function and then graph them. Use TaylorPlot to plot the Taylor polynomials for f(x)=cos(x). (Because of the answer to the previous question, use nSkip=2.) A good plotting interval is xmin = -9 and xmax = 9. Complete and Enter the commands below.

```
> xmin := -9:  xmax := 9: ymin := ???: ymax := ???:
> TaylorPlot(cos(x), x = xmin .. xmax, n = 0 .. 18, 2, y=ymin..ymax);
```

What is the relationship between the graphs of p(1,x), p(2,x),..., and the graph of cos(x)?. Is there an n for which p(n,x)=cos(x)? Place your answers below.
Answer:

If you conjectured the existence of an n for which p(n,x)=cos(x), then be sure to "double check" your claim by increasing the size of the x-interval in TaylorPlot and executing the command again. Does your claimed equality still remain valid? Place your answers below.
Answer:

As done for the exponential function in the previous section, we can generate a graph of the approximation errors by using the command TaylorErrorPlot. Use the command to plot the errors for f(x)=cos(x) for 0<=n<=12. (Use nSkip=2 for the same reasons as before.) Enter the following command.

```
> TaylorErrorPlot(cos(x), x = xmin .. xmax, n = 0 .. 18, 2);
```

The error curves can be used to approximate values of cos(x) to specified accuracies. As an example, use the error curves to find the smallest value of n such that p(n,2) approximates cos(2) to within an error of at most 0.01. Display this approximation p(n,2) and check its value against the "exact" value of cos(2). Place your computations and answer below.

```
>
```

Answer:

¤ 5. Your Choice.

Pick a function f(x) of your own--you will analyze its Taylor polynomials below. Enter your function definition below.

• Advice. You must pick a function f(x) which is defined and (many times) differentiable at x=0. Moreover, you should not pick just a polynomial (you already know what happens for a polynomial!), but you should also not pick too complicated a function (or else computations will take very long). One good choice is (x+2)^(1/3) over the interval -3.5 <= x <= 4.5. Another class of good choices are of the form (p(x))/(q(x)) where p(x) and q(x) are specific polynomials in x.

```
> f := x -> ???;
```

Use the command TaylorPoly to display some Taylor polynomials p(n,x) for f(x). Complete and Enter the following command.

```
> p := proc(n,x) TaylorPoly(f(x), x, n, x); end:
> p( ??? ,x);
```

Use the command TaylorPlot to show convergence (or divergence) of the Taylor polynomials to the original function f(x). Complete and Enter your commands below.

```
> plot(f);
> xmin := ???: xmax := ???: ymin := ???: ymax := ???:
> TaylorPlot(f(x), x = xmin .. xmax, n = 0 .. 18, 1, y = ymin .. ymax);
```

What is the relationship between the graphs of p(1,x), p(2,x),..., and the graph of your function f(x)?. In particular, for what values of x does it appear that p(n,x) converges to f(x) as n gets large? For what values of x does it appear the p(n,x) does not converge to f(x) as n gets large? Place your answers below.

Answer:

Use the command TaylorErrorPlot to display the error curves obtained when approximating f(x) by its various Taylor polynomials. Pick a value of x0 in your interval (xmin, xmax) and determine which polynomial p(n,x0) you will need to approximate f(x0) to within an error of .01. Enter your calculations and discussion below.

```
>
```

Discussion:

□

LAB 25: Taylor Polynomials II

□ NAMES:

□ PURPOSE
In this laboratory experiment we continue our investigation of the Taylor polynomial approximations to functions. The questions we concentrate on are the following:

Do the Taylor polynomials p(n,x) for a function f(x) converge to f(x) for all x as n gets large? If not, then what can be said about the set of x for which such convergence does take place?

Are there more efficient ways of obtaining Taylor polynomials than calculation of the defining formulas for the coefficients (which can be unpleasant to use for a complicated function)?

In the process of investigating these questions, we will discover polynomial approximations for the natural logarithm and the inverse tangent.

□ PREPARATION
Study Sections 11.3 and 11.4, pages 724 - 745, in Calculus: Modeling and Application. It would be helpful to review the previous laboratory experiment, Taylor Polynomials I.

□ LABORATORY REPORT INSTRUCTIONS
For this laboratory you are not required to submit a formal report. Rather you are asked to submit a copy of this Worksheet with the questions answered in the indicated locations. Only one submission will be made. Write in complete sentences and be sure the sentences connect together in coherent ways.

□ Maple Procedures.

```
> with(lab25);
> # If an Error is the output of the above line, see your lab instructor.
```

The commands TaylorPoly, TaylorPlot, and TaylorErrorPlot commands are the same as in the previous laboratory, Taylor Polynomials I, and will be recalled and described when needed during the current lab.

□ PRELIMINARIES

¤ On the Calculation of Taylor Polynomials by Maple.

For your convenience we recall the definition of the Taylor Polynomials for a function f(x) about the value a=0. The n-th Taylor Polynomial is defined to be

p(n,x) = a0 + a1*x + a2*x^2 + ... + an*x^n

where the coefficients a0, a1, a2, ..., an are defined by the formula

ak = (D@@k)(f)(0)/k!

In the previous lab we saw that as n gets large the polynomials p(n,x) often appear to converge to f(x), i.e., the graphs of the polynomials become better and better approximations to the graph of the original function. We wish to investigate this property further in this lab.

We have a command TaylorPoly to easily compute the Taylor polynomials for a specified function. Obtain information about this command in the usual fashion. Enter your command below.

```
>
```

As an example of the use of TaylorPoly, the following commands define p(n,x) to be the n-th Taylor polynomial for f(x)=(2+x)^(1/3). Enter these commands, and then

(1) display the 6-th Taylor polynomial for f(x), and

(2) calculate the value of the 6-th Taylor polynomial for f(x) at x=1.5. Enter your commands below.

```
> f := x -> ???;
> p := proc(n,x) TaylorPoly(f(t), t, n, x); end:
```

We must raise an issue of efficiency. Suppose you want to repeatedly use the 6-th Taylor polynomial for f(x). In such a case using p(6,x) is extremely inefficient (i.e., sloooow) because it requires the re-computation of the Taylor coefficients every time it is executed! We would like to execute p(6,x) just once, and store the result as a new function, p6(x). This is done in the following way. Enter the following definition for p6(x).

```
> p6 := unapply(p(6,x),x);
```

In case you do not believe there is a significant difference in speed, the following comparison should be enlightening. Execute the following two commands.

```
> st := time( ): seq(p(6,i/100),i=-100..100); time( )-st;

> st := time( ): seq(p6(i/100),i=-100..100); time( )-st;
```

Which method do you prefer!! ______________ The difference is even more dramatic for larger values of n.

□ PROJECT

¤ 1. The Function 1/(1-x).

Use the command TaylorPoly to find the Taylor polynomials of degrees 0 through 4 for the function f(x) = 1/(1-x). Enter your commands below.

```
> f := ???;
> TaylorPoly( ??? );
```

From these results, guess the general form for the Taylor polynomial p(n,x): Place your conjecture below.

- Conjecture: p(n,x) = ???

Check your result by calculating p(5,x) with the TaylorPoly command. Enter your command below.

```
>
```

If your answer does not agree with your conjecture, change your conjecture and check your new statement by calculating p(6,x). Enter your commands below (if necessary).

```
>
```

Earlier in this course you've seen a closed form expression for the addition of all the terms in your conjecture, i.e., an expression "?" without a summation sign or an ellipsis(...) such that:

p(n,x) = "?"

Do you remember this result? Whether or not you remember, let's re-derive it (it's important enough to repeat!). Begin by placing your polynomial expression for p(n,x) into the formula below; the resulting equation can then be solved for p(n,x), and produces the desired mystery expression "?". Place your computations below.

```
p(n,x) - x*p(n,x) = (     ) - x*(     )
                  = (     ) - (     )
                  = (     )
Hence,     p(n,x) = (     ) / (     )
```

This is one of the few cases where we have a simple formula for the n-th Taylor polynomial p(n,x) for a function f(x). In particular, it allows us to determine those x values for which the Taylor polynomials p(n,x) converge to f(x) as n gets large. Answer the following three questions.

- For what values of x do we have

```
> Display(1);
```

$$\lim_{n \to \infty} p(n, x) = f(x)$$

Answer:

• For what x value is our expression for p(n,x) invalid? What happens as n gets large in this case?
Answer:

• For what values of x do we have p(n,x) not converging as n gets large?
Answer:

Let's graphically check your answers. First take your formula for p(n,x) and put it into a Maple function definition. Complete and enter the following commands.

```
> p := proc(n,t) ??? end:
```

Now use the command TaylorPlot to plot the function f(x) = 1/(1-x) along with the Taylor polynomials p(n,x) of degree 0 through 9. (Note: As in the previous Taylor polynomial lab, you may need to experiment to find the best plot region a <= x <= b, c <= y <= d.) Complete and enter the following command.

```
> TaylorPlot([f,p],x=???..???,n=0..15,1,y=???..???);
```

Do the pictures justify the claims you made about the convergence properties of the p(n,x)? Place your answer below.
Answer:

Since we have a simple formula for p(n,x), even for large n we can painlessly plot the Taylor polynomials. Use the TaylorPlot command to "zoom in" on an interesting part of the graph and plot some Taylor polynomials for large values of n (say around n=100). Do these plots continue to confirm the convergence properties you claimed earlier? Enter your commands below.

```
> TaylorPlot ???
```

From the work done in this section--both visual display and analytic computation--you should have a good understanding of the Taylor polynomials p(n,x) for the function f(x)=1/(1-x). Of course, the thoroughness of our analysis was made possible by the simple closed form expressions that we had for p(n,x), something that rarely exists with other functions.

However..., might it be possible to use our knowledge of the behavior of the Taylor polynomials for f(x)=1/(1-x) to obtain Taylor polynomial information about other functions?

The answer is an emphatic YES, as we shall now show in the rest of this Worksheet.

¤ 2. The Functions 1/(1+x) and 1/(1+x^2).

We begin this section with a summary of the major facts determined in the previous section. Complete the following statements.

```
> Display(2);
```

$$\lim_{n \to \infty} p(n, x) = \lim_{n \to \infty} ??? + ??? + ??? + \ldots + ???$$

1/(1-x) = limit(p(n,x), n=infinity) = ??? + ??? + ??? + ...
Formula is valid for ?? < x < ?? (interval of convergence)
Formula fails for x < ?? or ?? < x
Formula also fails for x = ?? or x = ?? (endpoints)

Copy this summary, Paste it below, and replace EVERY x you see with a negative x. Then algebraically simplify the last three lines of inequalities to convert the -x back into an x. Place your answers below.
Answer:

Notice what you have obtained: the function g(x)=1/(1+x) has now been expressed as the limit of a sequence of polynomials of the form

q(x,n) = a0 + a1*x + a2*x^2 + ... +an*x^n,

at least for all values of x in some "interval of convergence" centered at 0. Fact: there is only one such set of polynomials for any function--the Taylor polynomials! (We will not attempt to prove this fact, but hopefully it should seem like a reasonable result.) Hence, without any hard work, we have obtained the Taylor polynomials for g(x)=1/(1+x) and have determined exactly those values of x for which the polynomials converge to g(x).

Let's check our claimed set of Taylor polynomials against the polynomials produced by TaylorPoly. Use this command to generate the 8-th Taylor polynomial for g(x)=1/(1+x) and compare it with the polynomials obtained above. Do we have agreement? Enter your commands and answers below.

```
> g := ???;
> TaylorPoly ???
```

- (Optional) Use TaylorPlot to graphically study the convergence of the Taylor polynomials to 1/(1+x). In particular, do the convergence properties match the claims made above? (The nature of the convergence for the Taylor polynomials for 1/(1+x) is just a "mirror" image of the convergence for 1/(1-x). That is why we have made the computation optional.) Enter your commands below.

```
>
```

* * * * * * * * * * * *

Let's turn to another example. We start again with the summary of the Taylor polynomials for 1/(1-x) found at the beginning of the section. Copy this summary, Paste it below, and replace EVERY x you see with a negative x^2. Then algebraically simplify the last three lines of inequalities to be in terms of x, not -x^2. Place your answers below.
Answers:

Notice what you have obtained: the function h(x)=1/(1+x^2) has now been expressed as the limit of a sequence of polynomials of the form

r(x,n) = a0 + a1*x + a2*x^2 + ... +an*x^n,

at least for all values of x in some "interval of convergence" centered at 0. However, as we stated earlier, there is only one such set of polynomials for any function--the Taylor polynomials! Hence we have obtained the Taylor polynomials for h(x)=1/(1+x^2) and have determined exactly those values of x for which convergence to h(x) occurs.

Let's check our claimed set of Taylor polynomials against TaylorPoly. Use this command to generate the 8-th Taylor polynomial for g(x)=1/(1+x) and compare it with the polynomials obtained above. Do we have agreement? Enter your commands and answers below.

```
>
```

Use TaylorPlot to graphically study the convergence of the Taylor polynomials to 1/(1+x^2). In particular, do the convergence properties match the claims made above? Enter your commands below.

```
> TaylorPlot(h,x = ???..???, n = 0 ..16, 2, y=???..???);
```

¤ 3. The Function log(1+x).

We start this section by Copying the summary on the Taylor polynomials for 1/(1+x) (notice: plus x, not negative x) from the previous section. Place the copy below.

For convenience in subsequent work, change the variable x in the expressions to t, i.e., replace every occurrence of x with t. Carry out the replacements in the summary you Pasted above.

Now observe: the definite integral of 1/(1+t) from t=0 to t=x is log(1+x), i.e.,

```
> Display(3);
```

$$\int_0^x \frac{1}{1+t}\,dt = \ln(1+x)$$

- Now suppose that g(n,t) is the Taylor polynomial for 1/(1+t).

- Might it be true that the integral of g(n,t) from t=0 to t=x is a Taylor polynomial for the integral of 1/(1+t) from t=0 to t=x, i.e., for log(1+x)?

Let's test this conjecture with some examples.

First we enter the Taylor polynomials q(n,t) for 1/(1+t) as Maple functions; this is most quickly done via the TaylorPoly command. Enter the following command.

```
> q := proc(n,t) TaylorPoly(1/(1+x),x,n,t) end:
```

We now integrate q(n,t) from t=0 to t=x, generating what we hope will be a Taylor polynomial for the function log(1+x). As an example, integrate q(6,t) from t=0 to t=x. Enter your command below.

```
> int ???
```

If our conjecture is true, the polynomial just obtained should be the 7-th degree Taylor polynomial for log(1+x). This is easy to check using TaylorPoly: let's generate the 7-th degree Taylor polynomial for log(1+x) and compare with the polynomial above. Do they agree? Enter your commands and answers below.

```
> TaylorPoly ???
```

Answer:

If the polynomial just produced does not match the one obtained by integrating q(6,t), then you need to check your work.

We let Q(n,x) denote the integral of q(n-1,t) from t=0 to t=x (we use n-1 since integration will raise the degree of the polynomial by 1). Then, as illustrated in the just completed calculations, what do we claim to be true about the Q(n,x)? Place your answer below.

Answer:

Define Q(n,x) with a Maple function definition command. Then display Q(9,x). Does the result appear to be correct? Enter your definition below.

```
> Q := proc(n,x) ??? end:
> Q ???
```

We wish to use the Q(n,x) polynomials in a TaylorPlot command in order to determine the interval of convergence. However, as currently defined, Q(n,x) is not efficient for computational purposes. (Do you recall the discussion in the Preliminaries section concerning efficiency of computation?) Although there are ways to redefine the Q(n,x) to improve computational efficiency , the easiest solution is to let the command TaylorPlot internally generate the Taylor polynomials for log(1+x). As you should readily believe by now, these Taylor polynomials are merely the Q(n,x) obtained by integration.

We now execute TaylorPlot for log(1+x). For what values of x does convergence of the Taylor polynomials Q(n,x) to log(1+x) appear to take place? Enter your commands and answers below.

```
> TaylorPlot(log(1+x), x = ??? .. ???, n = 1 .. 31, 2, y = ??? .. ???);
```

Answer:

How does the interval of convergence for the Taylor polynomials Q(n,x) of log(1+x) compare with the interval of convergence for the Taylor polynomials q(n,t) of 1/(1+t)?
Answer:

The behavior of the Taylor polynomials Q(n,x) at 1 might not be totally clear. On the basis of the following TaylorPlot pictures, decide whether or not the Taylor polynomials for log(1+x) converge at x=1. Execute the following command, and then give your analysis.

```
> TaylorPlot( log(1+x), x = 0.9 .. 1.1, n = 11 .. 110, 11, y = 0.6 .. 0.8 );
```

Analysis:

For the sake of comparison, let's generate a similar set of pictures for the Taylor polynomials q(n,t) of 1/(1+t) at t=1. On the basis of the following TaylorPlot pictures, decide whether or not we have convergence at t=1. Does this claim match the conclusion we reached in Section 2? Execute the following command, and then give your analysis.

```
> TaylorPlot( 1/(1+t), t = 0.9 .. 1.1, n = 11 .. 110, 11, y = -0.1 .. 1.1 );
```

Analysis:

- Complete the summary below.

The Taylor polynomials q(n,t) for 1/(1+t) have the following interval of convergence:
??? (< or <=) t (< or <=) ???
The Taylor polynomials Q(n,x) for log(1+x) have the following interval of convergence:
??? (< or <=) x (< or <=) ???

¤ 4. The Function arctan(x).

We start this section by Copying the summary on the Taylor polynomials for 1/(1+x^2) from Section 2. Place the copy below.

For convenience in subsequent work, change the variable x in this summary to t, i.e., replace every occurrence of x with t. Carry out the replacements in the summary you Pasted above.

Now observe: the definite integral of 1/(1+t^2) from t=0 to t=x is arctan(x), i.e.,

```
> Display(4);
```

$$\int_0^x \frac{1}{1+t^2}\, dt = \arctan(x)$$

Let r(n,t) be the n-th Taylor polynomial for 1/(1+t^2). Then what do you think would be the significance of the definite integral of r(n,t) from t=0 to t=x? Place your conjecture below.
Conjecture:

Let's check your conjecture. First we enter the Taylor polynomials r(n,t) for 1/(1+t^2) as Maple functions; this is most quickly done via the TaylorPoly command. Enter the following command.

```
> r := proc(n,t) TaylorPoly(1/(1+x^2),x,n,t) end:
```

Now define R(n,x) to be the definite integral of r(n-1,t) from t=0 to t=x (why do we use n-1 instead of n?). Then display R(9,x). Complete and Enter your commands below.

```
> R := proc(n,x) ??? end:
> ???
```

If your conjecture is true, then what relationship should exist between the polynomial R(9,x) and the function arctan(x)? Place your answer below.
Answer:

Check whether or not this claim is true by using the command TaylorPoly. Did you obtain what you expected? Enter your computations and answer below.

```
>
```

Answer:

We wish to use the R(n,x) polynomials in a TaylorPlot command in order to determine the interval of convergence. However, as currently defined, R(n,x) is not efficient for computational purposes. Thus, as done in the previous section for log(1+x), we will let TaylorPlot internally generate the Taylor polynomials for arctan(x).

We now execute TaylorPlot for arctan(x). For what values of x does convergence of the Taylor polynomials to arctan(x) appear to take place? Enter your commands and answers below.

```
> TaylorPlot(arctan(x),x=??? .. ???, n = 1 .. 21, 2, y = ??? .. ???);
```

How does the interval of convergence for the polynomials R(n,x) compare with the interval of convergence for the Taylor polynomials r(n,t) of 1/(1+t2)? Place your answer below.
Answer:

The behavior of the polynomials R(n,x) at 1 and -1 might not be totally clear. On the basis of the following TaylorPlot pictures, decide whether or not the R(n,x) converge at x=1. Similar behavior would be seen at x = -1. Execute the following command, and then give your analysis.

```
> TaylorPlot( arctan(x), x = 0.9 .. 1.1, n = 11 .. 101, 10, y = 0.7 .. 0.9 );
```

Analysis:

For the sake of comparison, let's generate a similar set of pictures for the Taylor polynomials r(n,t) of 1/(1+t2) at t=1. On the basis of the following TaylorPlot pictures, decide whether or not we have convergence at t=1. Does this claim match the conclusion we reached in Section 2? Execute the following command, and then give your analysis.

```
> TaylorPlot( 1/(1+t^2), t = 0.9 .. 1.1, n = 11 .. 101, 10, y = -0.05 .. 1.2 );
```

Analysis:

- Complete the summary below.

The Taylor polynomials r(n,t) for 1/(1+t^2) have the following interval of convergence:
??? (< or <=) t (< or <=) ???
The Taylor polynomials R(n,x) for arctan(x) have the following interval of convergence:
??? (< or <=) x (< or <=) ???

Consider the following conjecture: integrating of a set of Taylor polynomials will produce a new set of Taylor polynomials with the same interval of convergence. On the basis of your results for log(1+x) and arctan(x) in the last two sections, do you think this statement is True or False. If False, give a modified conjecture that you might be more inclined to believe. Place your answers below.
Answer:

¤ 5. Application to Approximations.

We wish to have one polynomial PAT(x) that will approximate all values of the arctan(x) for -0.4 < x < 0.4 to within an error of .000005 (i.e., five decimal place accuracy). For computational efficiency we desire PAT(x) to be of as small a degree as possible. Using the following TaylorErrorPlot command, determine a polynomial that we can use for PAT(x). Display the polynomial that you finally decide upon. Enter your commands and answers below.

```
> nMin := ???: nMax := ???: nSkip := ???:
> TaylorErrorPlot(arctan(x),x=???..???,n=nMin..nMax,nSkip,y=-???..???);
> PAT := t -> ???
```

Use PAT(x) to compute arctan(0.35) to five decimal place accuracy. Then check your answer against what Maple computes for arctan(0.35). Did you obtain five decimal place accuracy? Place your commands and answers below.

```
>
```

Answer:
□